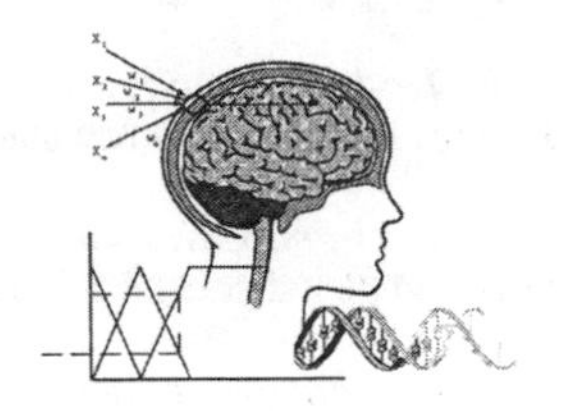

The CRC Press

International Series on Computational Intelligence

Series Editor
L.C. Jain, Ph.D., M.E., B.E. (Hons), Fellow I.E. (Australia)

L.C. Jain, R.P. Johnson, Y. Takefuji, and L.A. Zadeh
Knowledge-Based Intelligent Techniques in Industry

L.C. Jain and C.W. de Silva
Intelligent Adaptive Control: Industrial Applications in the Applied Computational Intelligence Set

L.C. Jain and N.M. Martin
Fusion of Neural Networks, Fuzzy Systems, and Genetic Algorithms: Industrial Applications

H.-N. Teodorescu, A. Kandel, and L.C. Jain
Fuzzy and Neuro-Fuzzy Systems in Medicine

C.L. Karr and L.M. Freeman
Industrial Applications of Genetic Algorithms

L.C. Jain and B. Lazzerini
Knowledge-Based Intelligent Techniques in Character Recognition

L.C. Jain and V. Vemuri
Industrial Applications of Neural Networks

H.-N. Teodorescu, A. Kandel, and L.C. Jain
Soft Computing in Human-Related Sciences

B. Lazzerini, D. Dumitrescu, L.C. Jain, and A. Dumitrescu
Evolutionary Computing and Applications

B. Lazzerini, D. Dumitrescu, and L.C. Jain
Fuzzy Sets and Their Application to Clustering and Training

L.C. Jain, U. Halici, I. Hayashi, S.B. Lee, and S. Tsutsui
Intelligent Biometric Techniques in Fingerprint and Face Recognition

Z. Chen
Computational Intelligence for Decision Support

L.C. Jain
Evolution of Engineering and Information Systems and Their Applications

H.-N. Teodorescu and A. Kandel
Dynamic Fuzzy Systems and Chaos Applications

L. Medsker and L.C. Jain
Recurrent Neural Networks: Design and Applications

L.C. Jain and A.M. Fanelli
Recent Advances in Artifical Neural Networks: Design and Applications

M. Russo and L.C. Jain
Fuzzy Learning and Applications

J. Liu and J. Wu
Multi-Agent Robotic Systems

M. Kennedy, R. Rovatti, and G. Setti
Chaotic Electronics in Telecommunications

H.-N. Teodorescu and L.C. Jain
Intelligent Systems and Techniques in Rehabilitation Engineering

I. Baturone, A. Barriga, C. Jimenez-Fernandez, D. Lopez, and S. Sanchez-Solano
Microelectronics Design of Fuzzy Logic-Based Systems

T. Nishida
Dynamic Knowledge Interaction

C.L. Karr
Practical Applications of Computational Intelligence for Adaptive Control

MULTI-AGENT ROBOTIC SYSTEMS

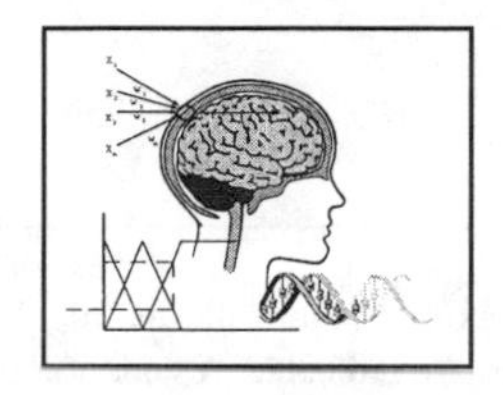

Jiming Liu
Jianbing Wu

CRC Press
Boca Raton London New York Washington, D.C.

Royalties from this book will be donated to Friends of the Earth and the World Wildlife Fund

Library of Congress Cataloging-in-Publication Data

Liu, Jiming, 1962-
Multi-Agent robotic systems / Jiming Liu, Jianbing Wu.
p. cm.—(International series on computational intelligence)
Includes bibliographical references and index.
ISBN 0-8493-2288-X (alk. paper)
1. Robots—Control systems. 2. Intelligent agents (Computer software) I. Wu, Jianbing. II. Title. III. CRC Press international series on computational intelligence

TJ211.35 .L58 2001 2001025397

Direct all inquiries to CRC Press LLC, 2000 N.W. Corporate Blvd., Boca Raton, Florida 33431.

Visit the CRC Press Web site at www.crcpress.com

International Standard Book Number 0-8493-2288-X
Library of Congress Card Number 2001025397
Printed in the United States of America 1 2 3 4 5 6 7 8 9 0
Printed on acid-free paper

Preface

Not everything that can be counted counts, and not everything that counts can be counted.[1]

An autonomous agent is a computational system that acquires and analyzes sensory data or external stimulus and executes behaviors that produce effects in the environment. It decides for itself how to relate sensory data to its behaviors in its efforts to attain certain goals. Such a system is able to deal with unpredictable problems, dynamically changing situations, poorly modeled environments, or conflicting constraints.

The motivation behind the research and development in multi-agent robotic systems comes from the fact that the decentralized multi-robot approach has a number of advantages over traditional single complex robotic systems approaches. Distributed robots can readily exhibit the characteristics of structural flexibility,

[1]A sign hanging in Albert Einstein's office at Princeton.

reliability through redundancy, simple hardware, adaptability, reconfigurability, and maintainability. The robots can interact with their local environments in the course of collective problem-solving. Responding to different local constraints received from their task environments, they may select and exhibit different behavior patterns, such as avoidance, following, aggregation, dispersion, homing, and wandering. These behaviors are precisely controlled through an array of parameters (such as motion direction, timing, lifespan, age, etc.), which may be carefully predefined or dynamically acquired by the robots based on certain computational mechanisms.

In order to successfully develop multi-agent robotic systems, the key methodological issues must be carefully examined. At the same time, the underlying computational models and techniques for multi-agent systems engineering must be thoroughly understood. As a companion to *Autonomous Agents and Multi-Agent Systems* by Jiming Liu, World Scientific Publishing, 2001, this book is about learning, adaptation, and self-organization in decentralized autonomous robots. The aims of the book are to provide a guided tour of the pioneering work and the major technical issues in multi-agent robotics research (Part I), and to give an in-depth discussion on the computational mechanisms for behavior engineering in a group of autonomous robots (Parts II to IV). Through a systematic examination, we aim to understand the interrelationships between the autonomy of individual robots and the emerged global behavior properties of the group in performing a cooperative task. Toward this end, we describe the essential building blocks in the architecture of autonomous mobile robots with respect to the requirements on local behavior conditioning and group behavior evolution.

This book has a number of specially designed features to enable readers to understand the topics presented and to apply the techniques demonstrated. These features are:

1. The contents have a balanced emphasis on the recent and pioneering work, the theoretical/computational aspects, and the experimental/practical issues of multi-agent robotic systems.

2. The materials are structured in a systematic, coherent manner with implementation details, comprehensive examples, and case studies.

3. Many graphical illustrations are inserted to explain and highlight the important notions and principles.

4. MATLAB toolboxes for multi-agent robotics research, experimentation, and learning are provided, which are available for free download from `http://www.crcpress.com/us/ElectronicProducts/downandup.asp?mscssid=`.

After reading this book, we hope that readers will be able to appreciate the strengths and usefulness of various approaches in the development and application of multi-agent robotic systems. They will be able to thoroughly understand the following five issues:

1. Why and how do we develop and experimentally test the computational mechanisms for learning and evolving sensory-motor control behaviors in autonomous robots.

2. How do we enable group robots to converge to a finite number of desirable task states through group learning.

3. What are the effects of the local learning mechanisms on the emergence of global behaviors.

4. How do we design and develop evolutionary algorithm-based behavior learning mechanisms for the optimal emergence of (group) behaviors.

5. How do we use decentralized, self-organizing autonomous robots to perform cooperative tasks in an unknown environment.

Jiming Liu
Jianbing Wu
Summer 2001

Acknowledgements

First, we would like to thank the pioneers and fellow researchers as well as the authors of books, papers, and articles in the fields of intelligent systems, autonomous agents and multi-agent systems, and robotics for providing us with insights, ideas, and materials in preparing the surveys in this book.

Second, we wish to offer our thanks to our home institutions, Hong Kong Baptist University (HKBU) and University of Calgary, for providing us with pleasant working environments that enable us to pursue our research. Without these environments, it could take longer for us to come up with this synthesis. Part of this book was written during Jiming's sabbatical leave in the Computer Science Department of Stanford University. Here we would like to thank Prof. Oussama Khatib for his kind invitation and also thank HKBU for granting this opportunity.

Third, we wish to acknowledge the research grants support provided by Hong Kong Baptist University under the scheme of Faculty Research Grants and by the Hong Kong Research Grants Council under the scheme of Earmarked Research Grants. Without that support, the theoretical and experimental studies reported in this book would not be possible.

In addition, we wish to thank X. Lai, K. P. Chow, and K. K. Hui for providing research assistance in some of the related studies. Our special thanks go to the editor of the series, Prof. Lakhmi C. Jain, for his vision and encouragement, and to Jerry Papke and Dawn Mesa of CRC Press for professionally managing and handling this book project. We also want to thank Madeline Leigh of CRC Press for carefully editing the draft version of this book.

And finally, to our families, to Meilee, Isabella, and Bernice from Jiming Liu, and to Audrey from Jianbing Wu, our heartfelt thanks for their love, inspiration, support, and encouragement.

Other Credits

Grateful acknowledgement is made to IEEE and ACM for permission to reuse some of our IEEE/ACM-copyrighted material in this book.

Portions of Jiming Liu et al., Learning coordinated maneuvers in complex environments: a SUMO experiment, *Proceedings of the IEEE/IEE Congress on Evolutionary Computation (CEC'99)*, pages 343-349, Washington, D.C., July 1999, are reused in Chapter 10. Portions of Jiming Liu and Jianbing Wu, Evolutionary group robots for collective world modeling, *Proceedings of the Third International Conference on Autonomous Agents (AGENTS'99)*, 1999, ACM Press, are reused in Chapters 12 and 13.

Figures 6.2, 6.11, 9.1-9.3, and 9.5-9.13 and Tables 7.1 and 9.1 are adopted, and Figures 6.15, 7.3(a), 8.6(a), 8.7(a), 8.8(a), 8.9(a), and 8.14(b) are modified, based on the figures and tables in Jiming Liu, Jianbing Wu, and Y. Y. Tang, On emergence of group behavior in a genetically controlled autonomous agent system, *Proceedings of the IEEE International Conference on Evolutionary Computation (ICEC'98)*, pages 470-475, Anchorage, May 1998.

Figures 11.1 and 11.4-11.10 are adopted, and Figures 11.2 and 11.3 are modified, based on the figures in Jiming Liu, Jianbing Wu, and Xun Lai, Analytical and experimental results on multiagent cooperative behavior evolution, *Proceedings of the the IEEE/IEE Congress on Evolutionary Computation (CEC'99)*, pages 1732-1739, Washington, D.C., July 1999.

Figures 12.3, 12.4, 12.12, 13.5-13.13, and 14.28 are adopted, and Figures 13.3 and 13.4 and Table 13.1 are modified, based on the figures and table in Jiming Liu, Jianbing Wu, and David A. Maluf, Evolutionary self-organization of an artificial potential field map with a group of autonomous robots, *Proceedings of the IEEE/IEE Congress on Evolutionary Computation (CEC'99)*, pages 350-357, Washington, D.C., July 1999.

Omissions of credit acknowledgement in this book, if any, will be corrected in future editions.

Jiming Liu
Jianbing Wu
Summer 2001

Contents

Part I

Motivation, Approaches, and Outstanding Issues

1

Why Multiple Robots?

So, then, to know a substance or an idea we must doubt it, and thus, doubting it, come to perceive the qualities it possesses in its finite state, which are truly "in the thing itself," or "of the thing itself," or of something or nothing. If this is clear, we can leave epistemology for the moment.[1]

Woody Allen

The field of distributed and cooperative robotics has its origins in the late 1980s, when several researchers began investigating issues in multiple mobile robot systems [AMI89, FN87]. Prior to this time, research had concentrated on either single-robot systems or distributed problem-solving systems that did not involve robotic components [Par00]. Since then, the field has grown dramatically, with a much wider variety of topics addressed.

[1] *Getting Even*, W.H. Allen & Co. Ltd., London, 1973, p 29.

1.1 Advantages

The use of multiple robots is often suggested to have many advantages over single-robot systems [BMF$^+$00, CFK97, DJMW96]. Cooperating robots have the potential to accomplish some tasks more efficiently than a single robot. Fox et al. [FBKT99] have demonstrated that multiple robots can localize themselves faster and more accurately if they exchange information about their positions whenever they sense each other. Furthermore, using several low-cost robots introduces redundancy and therefore is more fault-tolerant than having only one powerful and expensive robot. Generally speaking, a multi-robot system has the following remarkable properties:

- a larger range of task domains
- greater efficiency
- improved system performance
- fault tolerance
- robustness
- lower economic cost
- ease of development
- distributed sensing and action
- inherent parallelism
- insight into social and life sciences

1.2 Major Themes

In developing a multi-robot system, one of the primary concerns is how to enable individual robots to automatically program task-handling behaviors adaptive to the dynamic changes in their task environments. Several researchers have started to address the issue of multiple autonomous robot cooperation. Mataric [Mat94a, Mat94d] has developed a group behavior learning method in which heterogeneous reward function-based reinforcement learning is applied to associate the foraging subset (i.e., summation and/or switching) of six basic behaviors with triggering conditions. Fukuda and Iritani have proposed a mechanism for emerging group cooperative behaviors among decentralized autonomous robotic systems, called CEBOT (i.e., Cellular Robots) [FI95]. Their work simulates the emergence of group behaviors based on a globally stable attractor and the generation of new group behaviors based on bifurcation-generated new attractors.

Cao et al. [CFKM95] have highlighted five major themes in the mobile robot group studies; namely, group control architecture (e.g., decentralization and differentiation), resource conflict resolution (e.g., space sharing), the origin of cooperation (e.g., genetically determined social behavior vs. interaction-based cooperative behavior [Mat94c]), learning (e.g., control parameter tuning for desired cooperation), and geometric problem solving (e.g., geometric pattern formation). While the methodology for behavior engineering of a single autonomous robot may be simply defined as the stages of target behavior analysis, architecture design, implementation, training, and behavior assessment [CDB96], the emergence of complex robot group behaviors remains an open problem.

Multi-robot system design is challenging because the performance in such a system depends significantly on issues that arise from the interactions between robots [Bal98]. These interactions complicate development since they are not obvious in the hardware or software design but only emerge in an operating group. It is very hard, or even impossible, to model the group behaviors and design centralized controllers in a top-down manner for robot teams in unknown, unconstructed, or dynamic environments. Automatic methods for matching multi-robot configurations to tasks do not yet exist. Cooperation and robot-robot interference, for instance, are not considerations for a single robot but are crucial in multi-robot systems.

1.3 Agents and Multi-Agent Systems

Maes [Mae95] defines an agent as "a computational system that tries to fulfill a set of goals in a complex, dynamic environment." It can sense the environment through its sensors and act upon the environment using its actuators. Depending on the types of environment it inhabits, an agent can take many different forms. Agents inhabiting the physical world are typically robots. Maes proposes that an agent's goals can have many different manifestations: they can be *end goals* or particular states the agent tries to achieve, they can be a selective reinforcement or reward that the agent attempts to maximize, they can be internal needs or motivations that the agent has to keep within certain viability zones, and so on.

Following the considerations in [WJ95], a weak and a strong notion of agency can be distinguished. According to the weak notion, an agent displays the following properties:

- autonomy
- reactivity
- proactiveness

According to the more specific and strong notion, additional properties or mental attitudes, such as the following, are used to characterize an agent:

- belief, knowledge, etc. (describing information states)

- intention, commitment, etc. (describing deliberative states)
- desire, goal, etc. (describing motivational states)

As Maes [Mae95] states,

> An agent is called autonomous if it operates completely autonomously – that is, if it decides itself how to relate its sensory data to motor commands in such a way that its goals are attended to successfully.

An agent is said to be adaptive if it is able to improve its goal-achieving competence over time. Autonomous agents constitute a new approach to the study of artificial intelligence, which is highly inspired by biology, in particular ethology, the study of animal behavior. An autonomous agent approach is appropriate for the class of problems that require a system to autonomously achieve several goals in a dynamic, complex, and unpredictable environment.

In recent years, a rapidly growing interest has been shown in systems composed of several interacting autonomous agents instead of only a single agent. Weiss and Dillenbourg [WD99] propose at least three reasons for this interest in multi-agent systems:

1. They are applicable in many domains that cannot be handled by centralized systems.
2. They reflect the insight gained in the past decade in disciplines like artificial intelligence, psychology, and sociology that "intelligence and interaction are deeply and inevitably coupled to each other."
3. A solid platform of computer and network technology for realizing complex multi-agent systems is now available.

The bibliography in [JD87] lists earlier related work in multi-agent systems. A description of more recent work can be found in [Wei99].

1.4 Multi-Agent Robotics

We can make several interesting observations from natural multi-agent societies. For example, ant colonies have workers, soldiers, and a queen; hundreds of ants can shoulder a dead earthworm cooperatively from one place to another [Bal97]. Why does specialization occur? Are individuals born with skills and physical attributes? It may not be straightforward to answer these questions bounded only in natural systems, but we can investigate the issues in artificial societies – multi-agent robotic systems – and gain some insights from observations.

Some typical examples of multi-agent robotic systems are given in Table 1.1. As Arkin suggests [Ark98], CEBOT [FN87] may be regarded as the first among all such systems. It is a *cellular robotic system* with small robots that can dock together to produce a larger robot. The CEBOT research utilizes an architecture that has multiple parallel behaviors integrated by *vector summation*. The motivation of MARS [FMSA99] is to design lifelike robotics systems. Multi-Agent Robotic Systems research [Bal97] concerns learning in multi-robot teams, e.g., finding which conditions lead to the diversity in a learning team and how to measure it. Several tasks for Multi-Agent Robotic Systems, including formation, foraging, and soccer, are investigated. The goal of Biorobotics and Collective Intelligence research [PS98] at the University of Zurich is to model and learn from biological systems, such as the navigation behavior of insects or the walking dynamics of humans, and explore the emergence of structures in a group of interacting agents. Nicoud and his group [FGM+98] study the different ways of using autonomous robot teams to efficiently fulfill predefined missions. They define the problems of coordinating robot activities. Inspired by the collective intelligence demonstrated by social insects, they focus on the robot-robot and robot-environment interactions leading to robust, goal-oriented, and perhaps emergent group behaviors. ACTRESS (ACTor-based Robots and Equipments Synthesis System) [AMI89] is a multi-robot system designed for heterogeneous applications that focuses on communication issues. Normally, the robots act independently; but if a need arises, they negotiate with other robots to form a cooperative group to handle a specific task. Mataric [Mat92c, Mat93] has created behaviors for multi-robot systems using a subsumption style architecture [Bro86]. She has created homing, aggregation, dispersion, following, and wandering behaviors, and used them in a foraging task.

Project	Researcher	Organization	Website
CEBOT, MARS	T. Fukuda	Nagoya University, Japan	http://www.mein.nagoya-u.ac.jp/
Multi-Agent Robotic Systems	R. C. Arkin	Georgia Institute of Technology, USA	http://www.cc.gatech.edu/aimosaic/robot-lab/research/multi-agent.html
Multirobot Systems	M. J. Mataric	University of Southern California, USA	http://www-robotics.usc.edu/∼maja/group.html
Biorobotics	R. Pfeifer	University of Zurich, Switzerland	http://www.ifi.unizh.ch/groups/ailab/
Collective Robotics	J.-D. Nicoud	Swiss Federal Institute of Technology, Switzerland	http://diwww.epfl.ch/lami/
ACTRESS	H. Asama	RIKEN, Japan	http://celultra.riken.go.jp/∼asama/

TABLE 1.1. Some representative multi-agent robotics studies.

Particularly challenging tasks for multi-agent robotics are those that are inherently cooperative. Research themes that have been so far studied include:

- multi-robot path planning [YFO+00]
- traffic control [PY90]
- formation generation [AOS89]
- formation keeping and control [BH00, BA98, Wan89]
- target tracking [PT00, PM00]

- multi-robot docking [MMHM00]
- box-pushing [MC92, MNS95]
- foraging [Mat94a, Mat94c]
- multi-robot soccer [MAAO+99, RV00, SV98]
- exploration [BMF+00, SAB+00]
- localization [FBKT99, FBKT00]
- collision avoidance [FAAE98]
- transport [IOH98, KZ97]

Multi-agent robotic research is growing so rapidly that it is becoming very difficult to keep track of what is going on. For a reference, here are some journals that from time to time publish articles on multi-agent robotics:

- *Autonomous Robots:* A recent issue of the journal *Autonomous Robots* was dedicated to *colonies* of robots
- *Adaptive Behavior:* This new multidisciplinary journal provides the first international forum for research on adaptive behavior in animals and autonomous, artificial systems
- *Robotics and Autonomous Systems:* Its primary goal is to capture the state-of-the-art in both symbolic and sensory-based robot control and learning in the context of autonomous systems
- Other major journals in this field include:
  - *Artificial Life and Robotics*
  - *Autonomous Agents and Multi-Agent Systems*
  - *IEEE Transactions on Robotics and Automation*

Besides these journals, there are a number of international conferences featuring the state-of-the-art in this as well as other related fields:

- *IEEE/RSJ International Conference on Intelligent Robots and Systems (IROS)*
- *IEEE International Conference on Robotics and Automation (ICRA)*
- *International Conference on Simulation of Adaptive Behavior (SAB)*
- *International Conference on Autonomous Agents*
- *International Conference on Multi-Agent Systems (ICMAS)*

- *International Symposium on Distributed Autonomous Robotic Systems (DARS)*
- *International Symposium on Artificial Life and Robotics*
- *International Joint Conference on Artificial Intelligence (IJCAI)*

2

Toward Cooperative Control

Partnership is an essential characteristic of sustainable communities. The cyclical exchanges of energy and resources in an ecosystem are sustained by pervasive cooperation. Indeed, we have seen that since the creation of the first nucleated cells over two billion years ago, life on Earth has proceeded through ever more intricate arrangements of cooperation and coevolution. Partnership – the tendency to associate, establish links, live inside one another, and cooperate – is one of the hallmarks of life.[1]

Fritjof Capra

The cooperation of robots in unknown settings poses a complex control problem. Solutions are required to guarantee an appropriate trade-off in task objectives within and among the robots. Centralized approaches to this problem are not

[1] *The Web of Life*, Harper Collins Publishers, Great Britain, 1996, p 293.

efficient or applicable due to their inherent limitations – for instance, the requirement of global knowledge about an environment and a precise design to consider all possible states [PM00]. Distributed approaches, on the other hand, are more appealing due to their properties of better scaling and reliability.

2.1 Cooperation-Related Research

An overview of approaches and issues in cooperative robotics can be found in [Ark98, CFK97, Mat95b]. Parker [Par99] has demonstrated multi-robot target observation using the ALLIANCE architecture [Par94], where action selection consists of inhibition (through motivational behaviors). As opposed to ALLIANCE, Pirjanian and Mataric [PM00] have developed an approach to multi-robot coordination in the context of cooperative target acquisition. Their approach is based on multiple objective behavior coordination extended to multiple cooperative robots. They have provided a mechanism for distributed command fusion across a group of robots to pursue multiple goals in parallel. The mechanism enables each robot to select actions that not only benefit itself but also benefit the group as a whole. Hirata et al. [HKA+99] have proposed a decentralized control algorithm for multiple robots to handle a single object in coordination. The motion command is given to one of the robots, referred to as a leader; and the other robots, referred to as followers, estimate the motion of the leader from the motion of the object and handle the object based on the estimated reference.

Studies on cooperation in multi-agent robotic systems have benefited from a number of distinct fields such as social sciences, life sciences, and engineering. According to Cao et al. [CFK97], the disciplines that are most critical to the development of cooperative robotics include *distributed artificial intelligence*, *distributed systems*, and *biology*.

2.1.1 Distributed Artificial Intelligence

Grounded in traditional symbolic AI and social sciences, DAI is composed of two major areas of study: *Distributed Problem Solving* (DPS) and *Multi-Agent Systems* (MAS) [Ros93]. DPS considers how the task of solving a particular problem can be divided among agents that cooperate in dividing and sharing knowledge about the problem and its evolving solutions. One important assumption in DPS is that the agents are predisposed to cooperate. Cao et al. [CFK97] advocate DPS research on "developing frameworks for cooperative behavior between willing agents" rather than "developing frameworks to enforce cooperation between potentially incompatible agents." The MAS research studies the collective behavior of a group of possibly heterogeneous agents with potentially conflicting goals [CFK97]. Durfee et al. [DLC89] define MAS as "a loosely coupled network of agents that work together to solve problems that are beyond their individual capabilities."

2.1.2 *Distributed Systems*

The field of distributed systems is a natural source of ideas and solutions for studying multi-robot systems. Some researchers have noted that distributed computing can contribute to the theoretical foundations of cooperative robotics [CFK97, FMSA99]. In [CFK97], distributed control is considered as a *promising* framework for the cooperation of multiple robots, i.e., distributed control methods realize many advantages (flexibility, adaptability, and robustness etc.) when the population of robots increases. By noting the similarities with distributed computing, theories pertaining to deadlock, message passing, and resource allocation, and the combination of the above as primitives can be applied to cooperative robotics.

2.1.3 *Biology*

The majority of existing work in the field of cooperative robotics has cited biological systems as inspiration or justification [Bal94]. Well-known collective behaviors of ants, bees, and other eusocial insects provide striking proof that systems composed of simple agents can accomplish sophisticated tasks in the real world [CFK97]. Although the cognitive capabilities of these insects are very limited, the interactions between the agents, in which each individual obeys some simple rules, can result in the emergence of complex behaviors. Thus, rather than following the traditional AI that models robots as deliberative agents, some researchers in cooperative robotics have chosen to take a *bottom-up* approach in which individual agents are more like ants – they follow simple reactive rules [Mat94a, BHD94, SB93, DGF+91, BB99, DMC96]. The behavior of insect colonies can be generally characterized as self-organizing systems.

2.2 Learning, Evolution, and Adaptation

An important goal in the development of multi-agent robotic systems is to design a distributed control infrastructure to enable robots to perform their tasks over a problem-solving period without human supervision. These lifelong robotic systems must be capable of dealing with dynamic changes occurring over time, such as unpredictable changes in an environment or incremental variations in their own performance capabilities.

Learning, evolution, and adaptation endow an agent in a multi-agent system with the ability to improve its likelihood of survival within an environment through appropriate competition or cooperation with other agents. Learning is a strategy for an agent to adapt to its environment. Through its experience of interacting with the environment, the agent can form its cognition for the application of a specific behavior, incorporating certain aspects of the environment in its internal structure. On the other hand, evolution is considered as a strategy for a population of agents to adapt to the environment. Adaptation refers to an agent's learning by making adjustments with respect to its environment. As identified by Colombetti and

Dorigo [CD98], two kinds of adaptation that are relevant to multi-agent robotics are *evolutionary adaptation* and *ontogenetic adaptation*. The former concerns the way in which species adapt to environmental conditions through evolution, and the latter is the process by which an individual adapts to its environment during its lifetime. As far as behavior is concerned, ontogenetic adaptation is *a result of learning* . Adaptability allows agents to deal with noise in their internal and external sensors as well as inconsistencies in the behaviors of their environment and other agents.

In the opinion of Nolfi and Floreano [NF99], evolution and learning are two forms of biological adaptation that differ in space and time. Evolution is "a process of selective reproduction and substitution" based on the existence of a distributed population of individuals. Learning, on the other hand, is "a set of modifications taking place within each individual during its own lifetime." Evolution and learning operate on different time scales. Evolution is "a form of adaptation capable of capturing relatively slow environmental changes that might encompass several generations." Learning, on the other hand, "allows an individual to adapt to environmental changes that are unpredictable at the generational level." Learning may include a variety of mechanisms that produce adaptive changes in the individual during its lifetime, such as *physical development*, *neural maturation*, and *synaptic plasticity*.

Although evolution and learning are two distinct kinds of change that occur in two distinct types of entities, Parisi and Nolfi [PN96] argue that the two strategies may influence each other. The influence of evolution on learning is not surprising. Evolution causes the changes in a genotype.

> Each individual inherits a genome that is a cumulative result at the level of the individual of the past evolutionary changes that occur at the level of a population.

The individual's genome partially specifies a resulting phenotypic individual – it constrains how the individual will behave and what it will learn. The way is open for an influence of evolution on learning. On the other hand,

> Evolution can converge to a desired genome more quickly than if learning is absent, although it remains true that learned changes are not inherited. If evolution is unaided by learning, its chances of success are restricted to the case that the single desired genome suddenly emerges because of the chance factors operating at reproduction. Learning can accelerate the evolutionary process both when learning tasks are correlated with the fitness criterion and when random learning tasks are used.

From an evolutionary perspective, learning has several adaptive functions. It allows individuals to adapt to changes in the environment that occur in the lifespan of an individual or across a few generations. Learning supplements evolution, as it enables an individual to adapt to changes in the environment that happen too quickly to be tracked by evolution. In summary, learning can help and guide evolution [FU98, NF99].

Although the distinction between learning and adaptation is not always clear, Weiss [Wei96] has shown that multi-robot learning can usually be distinguished from multi-robot adaptation by the extent to which new behaviors and processes are generated. Typically, in multi-robot learning, new behaviors or behavior sequences are generated, or functions are learned, thus giving a robot team radically new capabilities. Frequently, the learning takes place in an initial phase, where performance during learning is not of importance. In multi-agent adaptation, the robot team exercises a control policy that gives reasonable results for the initial situation. The team is able to gradually improve its performance over time. The emphasis in multi-robot adaptation is the ability to change its control policy online – while the team is performing its mission – in response to changes in the environment or in the robot team.

In multi-agent robotics, evolutionary algorithms have been widely used to evolve controllers [HHC92, HHC+96, Har96, Har97, HHCM97]. Generally speaking, the controllers that become well adapted to environmental conditions during evolution may not perform well when the conditions are changed. Under these circumstances, it is necessary to carry out an additional evolutionary process, which, as Urzelai and Floreano [UF00] have stated, can take a long time. On the other hand, the integration of evolution and learning may offer a viable solution to this problem by providing richer adaptive dynamics than when parameters are entirely genetically determined.

2.3 Design of Multi-Robot Control

Finding the precise values for control parameters that lead to a desired cooperative behavior in multi-robot systems can be a difficult, time-consuming task for a human designer. Harvey et al. [HHC+96] point out that there are at least three major problems that a designer may encounter:

1. It is not clear how a robot control system should be decomposed.
2. The interactions between separate subsystems are not limited to directly visible connecting links; interactions are also mediated via the environment.
3. As system complexity grows, the number of potential interactions between the components of the system grows exponentially.

As Harvey [Har96] indicates, classical approaches to robotics have often assumed a primary decomposition into perception, planning, and action modules. Brooks [Bro86], on the other hand, acknowledges problems 2. and 3. in his subsumption architecture, and he advocates the careful design of a robot control system layer by layer by hand. An obvious alternative approach is to abandon

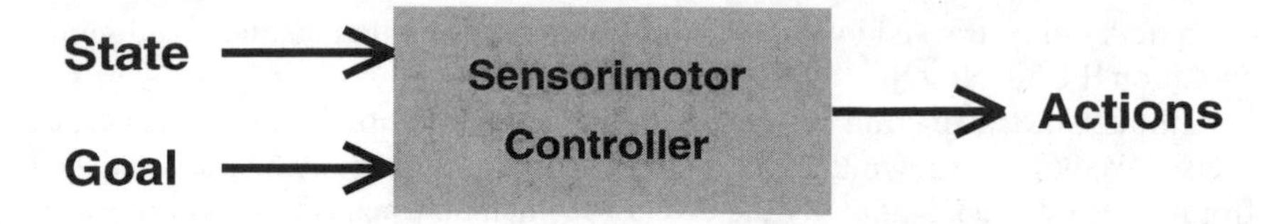

FIGURE 2.1. A general model of robotic control.

hand design and explicitly use evolutionary techniques to incrementally evolve complex robot control systems.

The control in a multi-agent robotic system determines its capacities to achieve tasks and to react to various events. The controllers of autonomous robots must possess both decision-making and reactive capabilities. And the robots must react to the events in a timely fashion. Figure 2.1 presents a general model of robot controllers. Generally speaking, a robot controller should demonstrate the following characteristics [Ark98, ACF+98, BH00]:

1. **Situatedness:** The robots are entities situated and surrounded by the real world. They do not operate upon abstract representations of reality, but rather upon reality itself.
2. **Embodiment:** Each robot has a physical presence (a body). This spatial reality has consequences in its dynamic interactions with the world (including other robots).
3. **Programmability:** A useful robotic system cannot be designed only for a single environment or task. It should be able to achieve multiple tasks described at an abstract level. Its functions should be easily combined according to the task to be executed.
4. **Autonomy and adaptability:** The robots should be able to carry out their actions and to refine or modify the task and their own behavior according to the current goal and execution context as perceived.
5. **Reactivity:** The robots have to take into account events with time bounds compatible with the correct and efficient achievement of their goals (including their own safety).
6. **Consistent behavior:** The reactions of the robots to events must be guided by the objectives of their tasks.
7. **Robustness:** The control architecture should be able to exploit the redundancy of the processing functions. Robustness will require the control to be decentralized to some extent.
8. **Extensibility:** Integration of new functions and definition of new tasks should be easy. Learning capabilities are important to consider here; the architecture should make learning possible.
9. **Scalability:** The approach should easily scale to any number of robots.
10. **Locality:** The behaviors should depend only on the local sensors of each robot.

11. **Flexibility:** The behaviors should be flexible to support many social patterns.
12. **Reliability:** The robot can act correctly in any given situation over time.

3
Approaches

Intelligence without Reason can be read as a statement that intelligence is an emergent property of certain complex systems – it sometimes arises without an easily identifiable reason for arising....

We are a long way from creating Artificial Intelligences that measure up to the standards of early ambitions for the field. It is a complex endeavor and we sometimes need to step back and question why we are proceeding in the direction we are going, and look around for other promising directions.[1]

Rodney A. Brooks

[1]Chapter 8: Intelligence without Reason, *Cambrian Intelligence*, The MIT Press, Cambridge, Massachusetts, p 185.

3.1 Behavior-Based Robotics

Mataric [Mat98] states that behavior-based robotics designs controllers for endowing robots with intelligent behavior, based on "a biologically inspired philosophy that favors parallel, decentralized architectures." It draws on the idea of providing the robots with a range of basic behaviors and letting the environment determine which behavior is more suitable as a response to a certain stimulus. Sukhatme and Mataric [SM00] define behaviors as "real-time processes that take inputs from sensors and/or other behaviors and send outputs to actuators and/or other behaviors." In behavior-based robotics, basic behaviors are fundamental units for control, reasoning, and learning. The environment plays a central role in activating a certain basic behavior at any given time. The behavior modules and the coordination mechanisms are usually designed through a trial-and-error process in which a designer progressively changes them and tests the resulting behavior in the environment. Extending the reactive and behavior-based approaches to a multi-agent domain will lead to completely distributed systems with no centralized controller.

Behavior-based robotics has been an active and popular approach to robot control in the multi-robot domain, allowing multi-robot systems to adapt to real-world environments. Behavior-based systems are praised for their robustness and simplicity of construction [Bro86, Mae89, Mat92a]. Based on Brooks' behavior-based subsumption architecture [Bro86], for example, Parker developed the ALLIANCE architecture [Par94] for controlling groups of heterogeneous robots and demonstrated it on a group of four physical robots performing puck manipulation and box-pushing. He divides tasks into subtasks, with groups of behaviors addressing each subtask. At the highest level, "mutually inhibitory motivational behaviors are designed to direct the overall behavior of a robot, which in turn activates lower-level behaviors to perform a subtask" [Par94]. Along with the typical sensor-based conditions that might trigger motivational behaviors, Parker adds impatience and acquiescence. Impatience increases if no other robot is tempting to solve the subtask associated with a motivational behavior, while acquiescence inhibits the behavior if the robot is not successful in the subtask. The combination of the ordinary conditions of impatience and acquiescence in a group enables the robots to cooperate in striving to achieve an overall task.

Balch [Bal97] takes motor schemas as an example of behavior-based robot control. Motor schemas are the reactive components in an Autonomous Robot Architecture (AuRA) [AB97]. AuRA's design integrates "deliberative planning at the top level with behavior-based motor control at the bottom." The lower levels are concerned with executing reactive behaviors. Individual motor schemas, or primitive behaviors, express separate goals or constraints for a task. For example, the schemas for a navigational task may involve avoiding obstacles and moving to a goal. Since schemas are independent, they can run concurrently, providing parallelism and efficiency. Motor schemas may be grouped to form more complex, emergent behaviors.

3.2 Collective Robotics

Ant colonies are able to collect objects (such as food or dead ants) and place them in particular places. The term collective behavior generically refers to "any behavior of a system having more than one agent" [CFK97]. Collective behaviors offer the possibility of enhanced task performance, increased task reliability, and decreased cost over traditional robotic systems.

Much work to date in collective robotics focuses on limited cases, such as flocking and foraging. Typical agents in such experiments either use manually built (non-learning) controllers [BA95, BHD94], or perform learning in simulated [Bal97, SP96] or relatively simple physical domains/environments [Mat94d, UAH98].

One way to generate robust collective behaviors is to apply biologically inspired adaptive algorithms at the team level. We believe that the integration of learning methods can contribute significantly to the design of a team of self-programming robots for some predefined tasks. In the past few years, reinforcement learning and genetic algorithms have been used to produce adaptive behaviors in the context of single-robot applications [FM96]. In multi-robot applications, where fitness is measured at the team level, robots will be faced with the credit assignment problem – deciding to what extent their behavior has contributed to the team's overall score [VG97]. Two ways for bypassing this problem have been proposed. The first is to integrate the global communication among teammates [Par95]. However, this is not a completely decentralized solution and does not match the above definition of biologically inspired robotics. Furthermore, depending on the environmental conditions, global communication is not always possible and tends to become a bottleneck when the size of a team increases. The second way is to measure individual performance for each robot instead of team performance [Mat96]. A drawback of this approach is that it forces a collective behavior to be the sum of individual behaviors, which is not necessarily the optimal strategy for a shared mission. Martinoli and Mondada [MM95] have implemented two biologically inspired collective behaviors in a group of miniature mobile robots. Martinoli et al. [MFM97] also provide a reliable setup for conducting bio-inspired experiments with real robots.

3.3 Evolutionary Robotics

Evolutionary computation has attracted attention from various research fields as a way of solving optimization problems [BHS97, Bre62, Mic92, Sch95, Sha94]. Robotics is one of such fields in which researchers have found many applications, ranging from control strategy synthesis to geometric motion planning [FAvN+96, HHC+96, HCH96, MC96]. As a result, a new wave in robotics has started to emerge; we may call it *evolutionary robotics*.

In evolutionary robotics, there are a number of representative issues; some are concerned with robot design, whereas others are concerned with planning and control. For instance, Murray and Louis [ML95] used a genetic algorithm to balance evolutionary design and human expertise in order to best develop robots that can learn specific tasks. Nolfi et al. [NFMM94], Steels [Ste94], and Xiao et al. [XMZT97] have applied genetic algorithms to various problems in robotics, such as design, behavior emergence, motion planning, and navigation. Harvey [Har92] has extended standard genetic algorithms for a finite search space with fixed-length genotypes to an open-ended evolution with variable-length genotypes.

Evolutionary robotics is a promising new approach to the development of multi-robot systems capable of reacting quickly and robustly in both real and simulated environments. In this discipline, algorithms inspired largely by biological evolution are used to automatically design sensorimotor control systems [NFMM94]. Although in the early days artificial evolution was mainly seen as a strategy to develop more complex robot controllers, today the field has become much more sophisticated and diversified. Floreano and Urzelai [FU00] have identified at least three approaches to evolutionary robotics: *automated engineering*, *artificial life*, and *synthetic biology/psychology*. These three approaches overlap with each other but still have quite different goals that eventually show up in the results obtained. *Automated engineering* is "about the application of artificial evolution for automatically developing algorithms and machines displaying complex abilities that are hard to program with conventional techniques;" *artificial life* is "about the evolution of artificial creatures that display lifelike properties;" and *synthetic biology/psychology* attempts to "understand the functioning of biological and psychological mechanisms by evolving those mechanisms in a robot put in conditions similar to those of animals under study."

In evolutionary robotics, the predominant class of systems for generating behaviors is Artificial Neural Networks (ANNs) [Hus98]. ANNs can be envisaged as simple nodes connected together by directional wires along which signals flow. The nodes perform an input-output mapping that is usually some sort of *sigmoid function*. Many people have applied ANNs in evolutionary robotics [Dye95, Har96, Har97, Har00, HHC+96, Sha97, Zie98]. Jakobi [Jak98b] has stated some specific reasons why ANNs are generally preferred:

1. By varying the properties and parameters of simple processing units used, different types of functionality can be achieved with the same type of network structure. This means that the same encoding schemes can be used independently of the functionality of the control system.

2. Using ANNs allows us to implement and test ideas from biology about how neural mechanisms for the generation of behavior may work. Also, it allows us to test and refine proposed biological models through the exploration of parameter spaces that may suggest new hypotheses.

3. There are other adaptive processes that we may use in conjunction with artificial evolution, such as various forms of supervised and unsupervised learning.
4. The behaviors that evolutionary robotics is concerned with at present are low-level behaviors, tightly coupled with the environment through simple, precise feedback loops. ANNs are ideal for this purpose.

Much of the evolutionary robotics work that has been undertaken to date concerns the evolution of fixed-architecture neural networks for the control of robots [FU00, NP95]. However, it is also possible to evolve neural networks for robot control whose size and morphology are under the control of an evolutionary process itself [Har97, JQ98]. Several systems have combined the power of neural controllers and genetic algorithms. Lewis et al. [LFB94] used genetic algorithm-based (GA) methods to evolve the weights for a neural controller in a robotic hexapod, named *Rodney*, rather than using traditional neural learning. Apart from neural networks, various types of control architectures have been evolved for robots (either simulated or real). The most common are techniques from genetic programming [KR92, LHL97a, LHF+97]. Other types of controller architectures that have been employed include classifier systems [DC94, DMC96, SG94].

Another important research direction in evolutionary robotics is to evolve and grow entire robots (e.g., the evolution of structures and body plans for LEGO robots) [FP97, LHL97b], in addition to evolving controllers given their morphologies. The significance of this work is that in robot manipulations, sometimes morphology can play an essential role. Depending on particular shapes, certain tasks may become much easier to perform than others.

Related to robot morphology evolution is the earlier work on evolving three-dimensional animated creatures. Ventrella [Ven94] has studied the possibility of emerging the structure and locomotion behaviors of an artificial creature using genetic algorithms. His system uses a model of specifically tailored qualitative forward dynamics to generate gravitational, inertial, momentum, frictional, and dampening effects. Sims [Sim94] has developed a system in which both the animated three-dimensional creatures' bodies (i.e., morphology) and their neural control systems (i.e., *virtual brains*) are genetically evolved and/or coevolved.

Mataric and Cliff [MC96] have identified several key problems with existing evolutionary robotics methods, such as simulator fidelity, evaluation time, and hardware robustness. *Embodied evolution* (EE) [WFP00], a new methodology for conducting evolutionary robotics, is proposed to ameliorate these particular concerns. Embodied evolution uses a population of physical robots that "evolve by reproducing with one another" in a task environment, where evaluation, selection, and reproduction are carried out by and between the robots in a distributed, asynchronous, and autonomous manner. However, some researchers have pointed out other questions [MC96, NFMM94] that need to be answered if evolutionary robotics is to progress beyond the proof-of-concept stage. One of the most urgent concerns is how evolved controllers can best be evaluated. If they are tested using real robots in the real world, then this has to be done in real time; and the evolution of complex behaviors will take a prohibitively long time. If controllers are tested using simulation, then the amount of modeling necessary to ensure that evolved controllers work on real robots may mean that the simulation is so complex to design and so computationally expensive that all potential speed advantages over real-world evaluation are lost.

3.4 Inspiration from Biology and Sociology

As pointed out by Colombetti and Dorigo [CD98], the view that animal behavior is best described as a number of interacting *innate motor patterns* has inspired the presently popular approaches to multi-robot control. These approaches apply some simple control rules from biological societies – particularly ants, bees, and birds – to the development of similar behaviors in cooperative robot systems. Work in this vein has demonstrated the ability of multi-robot teams in flocking, dispersing, aggregating, foraging, and following trails (see [Mat92b]). McFarland [McF94] has applied the dynamics of ecosystems to the development of multi-robot teams that emerge cooperation as a result of acting on selfish interests. Competition in multi-robot systems, similar to that in nature, is presently being studied in domains such as multi-robot soccer games (see [MAAO+99, SV98]). Other recently identified topics of relevance include the use of imitation to learn new behaviors, and the physical interconnectivity as demonstrated by insects, such as ants, to enable collective navigation over challenging terrains (see [BDT99, Mat94b, Sch99]).

As described by Parker [Par00], many cooperative robotics researchers have found it instructive to examine the social characteristics of insects and animals and to apply these findings to the design of multi-robot systems. The fields of *population biology* and *ecological modeling* are concerned with the large-scale *emergent* processes that govern systems of many entities interacting with one another. Population biology and ecological modeling in this context consider the "dynamics of the resultant ecosystem," in particular "how its long-term behavior depends on the interactions among the constituent entities" [WT99]. The field of *swarm intelligence* studies the systems that are modeled after social insect colonies [BDT99]. For instance, Beni [Ben88, BW89] has investigated cellular robotic systems where many simple agents occupy one- or two-dimensional environments to generate and self-organize patterns through nearest-neighbor interactions. Swarm intelligence may be viewed as ecological modeling where individual entities have extremely *limited computing capacity* and/or *action sets*, and where there are very few types of entities. The premise of this field is that the rich behavior of social insect colonies arises not from the sophistication of any individual entity in the colony, but from the interactions among those entities. The study of social insect colonies will also provide us with new insight into how to achieve learning in large-scale distributed systems [WT99].

In addition, ethological studies have shown that multi-agent societies offer significant advantages in performing community tasks. A wide range of animal social structures exists to support agent-agent interactions. For example, "unilevel organizations are found in schooling fish, hierarchical systems are found in baboon societies, and caste systems are typified by many insect colonies (such as bees)" [Ark98]. The relationships between these agents often determine the nature and type of communication essential for the social system to succeed. The converse also holds in that the communication abilities somewhat determine the most effective social organizations for a particular class of agents [Ark98]. McFarland [McF94] defines cooperative behavior as "the social behavior observed in higher animals (vertebrates) – cooperation is the result of interactions among selfish agents." Dautenhahn and Billard [BD97] have undertaken studies on *social robotics* where the emergence of a social behavior is achieved through the interaction of two robots via a simple imitative strategy. Dautenhahn and Nehaniv [DN98] have addressed the issues of social intelligence, communication, and body image. While interesting from an artificial life perspective, Dautenhahn and Nehaniv do not utilize an explicit communication protocol or architecture that directly supports formal social interactions. The social behaviors produced are primarily emergent and based on *semaphore communication*. Similar research by Pfeifer [Pfe98] demonstrates that very simple robotic entities can exhibit emergent social behaviors.

Furthermore, economies also provide examples of naturally occurring systems that demonstrate (more or less) collective intelligence. Both *empirical economics* (e.g., economic history, experimental economics) and *theoretical economics* (e.g., general equilibrium theory, theory of optimal taxation) provide rich literature on strategic situations where many parties interact [Bed92]. In fact, much of the entire field of economics is concerned with how to maximize certain world utilities, when there are some restrictions on individual agents and their interactions and in particular when we have limited freedom in setting the utility functions for those agents.

3.5 Summary

Behavior-based robotics, collective robotics, and evolutionary robotics have been inspired by biology, ethology, sociology and other related fields. The three approaches are interrelated. They all aim at generating complex, adaptive, and goal-driven group behaviors from simple local interactions among individuals in multi-agent robotic systems [Mat92b]. However, they also produce different forms of autonomy, adaptability, task complexity, and intelligence in multi-agent systems. For example, although behavior-based approaches are robust for many task environments, they are not necessarily adaptive. An evolutionary system can, on the other hand, improve the adaptability to the changes in a dynamic environment.

Figure 3.1 presents a comparative view of the three approaches along the dimensions of autonomy and task performance.

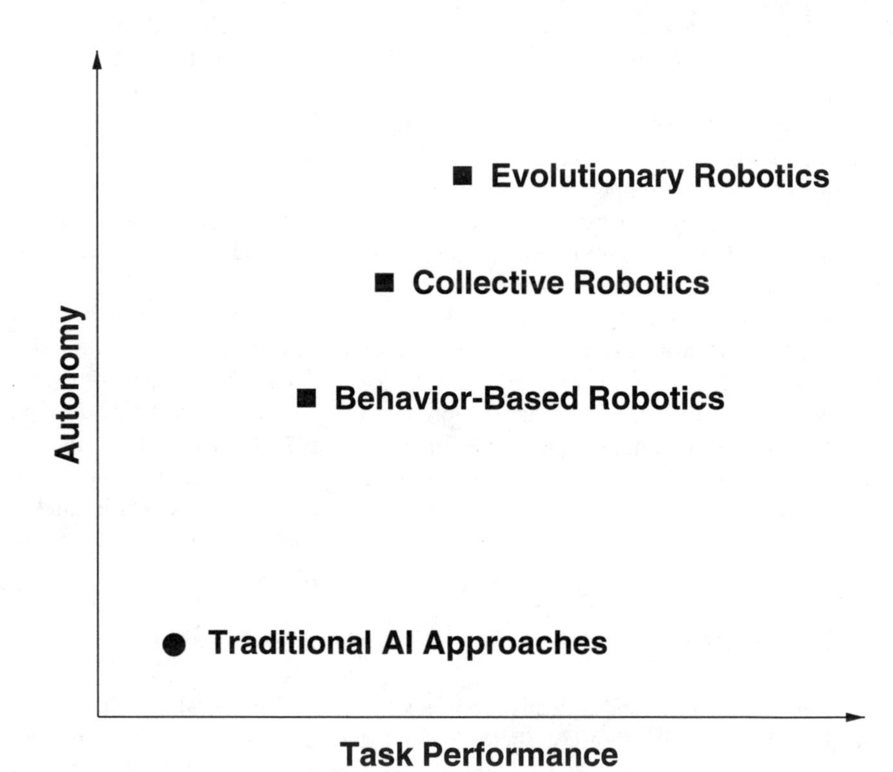

FIGURE 3.1. Approaches in multi-agent robotics research focusing on various objectives.

4
Models and Techniques

So even if we do find a complete set of basic laws, there will still be in the years ahead the intellectually challenging task of developing better approximation methods, so that we can make useful predictions of the probable outcomes in complicated and realistic situations.[1]

Stephen Hawking

4.1 Reinforcement Learning

Reinforcement learning (RL) is defined by Sutton and Barto in their book *Reinforcement Learning: An Introduction* [SB98] as "learning what to do – how

[1] *A Brief History of Time*, Bantam Press, Great Britain, 1988, p 187.

to map situations to actions – to maximize a numerical reward signal." In reinforcement learning, as in most forms of machine learning, "a learner is not told which actions may take, but instead must discover which actions yield the most reward by trying them." In most cases, actions may affect not only the immediate reward but also the next situation and through that, all subsequent rewards. In their opinion, "trial-and-error search and delayed reward are the two important distinguishing features of reinforcement learning." The trade-off between exploration and exploitation has been thought as a challenge arising in reinforcement learning, but not in other forms of learning. In order to gain a positive reward, a reinforcement-learning agent must prefer actions that it has tried in the past and found to be effective in producing a reward; but to discover new actions it has to try actions that it has not selected before. That means the agent has to exploit what it already knows in order to obtain a reward, and it also has to explore in order to make better action selections in the future.

Besides an agent and an environment, Sutton and Barto [SB98] have further identified four important elements of reinforcement learning: a *policy*, a *reward function*, a *value function*, and optionally, a *model of the environment*. Specifically,

> A policy defines the learning agent's way of behaving at a given time. Roughly speaking, a policy is a mapping from certain perceived states of the environment to actions to be taken when in those states. A reward function defines the goal in a reinforcement learning problem. It maps a perceived state (or state-action pair) of the environment to a single number, a reward, indicating the intrinsic desirability of the state. A reinforcement learning agent's sole objective is to maximize the total reward it receives in the long run. The reward function determines what are the good or bad events for the agent. A value function specifies what is good in the long run. The value of a state is the total amount of reward the agent can expect to accumulate in the future starting from that state. Whereas rewards determine the immediate, intrinsic desirability of environmental states, values indicate the long-term desirability of states after taking into account the states that are likely to follow and the rewards available in those states. A model of the environment is something that mimics the behavior of the environment. The agent can use such a model to predict how the environment will respond to its actions.

In multi-agent robotics, robotic agents may have many possible actions that they can take in response to a stimulus; and a policy determines which of the available actions the robots should undertake. Reinforcement is then applied based on "the results of that decision, and the policy is altered in a manner consistent with the outcome (reward or punishment)" [Ark98]. The ultimate goal is to "learn an optimal policy that chooses the best action for every set of possible inputs." The robots strive to improve their performance, finding suitable behaviors as they interact with their environment. This approach has the added benefit of allowing the agents to adapt to different environmental conditions. For this reason, reinforcement learning has become an attractive learning technique in multi-agent robotics [Bal98, KLM96].

4.1.1 Markov Decision Process

Mataric [Mat94a] has observed that in most computational models of reinforcement learning, the agent-environment interaction can be described as a *Markov decision process* (MDP). Her arguments are as follows:

1. The agent and the environment can be modeled as synchronized finite state automata.
2. The agent and the environment interact in discrete time intervals.
3. The agent can sense the state of the environment and use it to make actions.
4. After the agent acts, the environment makes a transition to a new state.
5. The agent receives a reward after performing an action.

4.1.2 Reinforcement Learning Algorithms

Mataric [Mat94a] also provides the following general form of reinforcement learning algorithms:

1. Initialize the learner's internal state I to I_0.
2. Repeat:
   (a) Observe current world state s.
   (b) Choose an action $a = F(I; s)$ using evaluation function F.
   (c) Execute action a.
   (d) Let r be an immediate reward for executing a in world state s.
   (e) Update internal state $I = U(I; s; a; r)$ using update function U.

The internal state I encodes the information the learning algorithm saves about the world, usually in the form of a table maintaining state and action data. The update function U adjusts the current state based on the received reinforcement and maps the current internal state, input, action, and reinforcement into a new internal state. The evaluation function F maps an internal state and an input into an action based on the information stored in the internal state. Different reinforcement learning algorithms vary in their definitions of U and F.

4.1.3 Temporal Differencing Techniques

As Mataric [Mat94a] indicates, the predominant methodology used in reinforcement learning is based on a class of temporal differencing (TD) techniques. All TD techniques deal with "assigning credit or blame to past actions by predicting long-term consequences of each action." Sutton's original formalization of temporal differencing deals with such predictions in a Markovian environment [Sut88].

4.1.4 Q-Learning

Balch [Bal97] has given a detailed description of Q-learning.

> It is a type of reinforcement learning in which the value of taking each possible action in each situation is represented as a utility function, $Q(s; a)$, where s is a state or situation and a is a possible action. If the function is properly computed, an agent can act optimally simply by looking up the best-valued action for the situation. The problem is to find $Q(s; a)$ that provides an optimal policy.

Watkins and Dayan [WD92] have developed an algorithm for determining $Q(s; a)$ that converges.

Lin [Lin93] has proposed a method of Q-learning where complex tasks can be learned hierarchically at several levels. The method decomposes a task into subtasks. A robot learns at the subtask level first, then at the task level. The overall rate of learning is increased if compared to monolithic learning. Similarities between Lin's decomposition and temporal sequencing for assemblages of motor schemas can be readily observed. Lin's subtasks or elementary skills correspond to behavior assemblages, while a high-level skill is a sequence of assemblages. Learning the elementary skills corresponds to tuning individual states or behavior assemblages, and learning at the high level is equivalent to learning the state transitions in a finite state automaton.

Since Q-learning is often used to deal with discrete actions and states, it may require a great deal of time and memory to learn the proper actions for all states. To reduce the complexity and generalize conventional Q-learning, Kim et al. [KSO+97] suggest a region-based credit assignment approach, where the regions are generated by applying an online convex clustering technique.

4.1.5 Multi-Agent Reinforcement Learning

To date, reinforcement learning is most often applied in single-robot systems [MB90, MC92]; but recent work indicates that multi-robot systems should benefit as well [Bal97, Mat97b]. As in other aspects of multi-robot systems design, when learning is extended from an individual to a team, new interactions will arise. Mataric [Mat94c] has investigated the learning of behavior-based multi-robot teams in foraging tasks. Her work focuses on developing specialized reinforcement functions for social learning.

> The overall reinforcement, R^t, for each robot is composed of separate components, D, O, and V. D indicates the progress of the robot toward its present goal. O corresponds to reinforcement if the present action is a repetition of another robot's behavior. V is a measure of vicarious reinforcement, which follows the reinforcement provided to other robots.

Mataric [Mat94a, Mat94d] has also proposed a reformulation of reinforcement learning using higher levels of abstraction (i.e., conditions, behaviors, heterogeneous reward functions, and progress estimators, instead of states, actions, and reinforcement) to enable robots to learn a composite foraging behavior.

Moreover, Nagayuki et al. [NID00] have developed a multi-agent reinforcement learning method based on the estimation of other agents' actions for solving a two-agent cooperation problem. Parker [Par94] uses standard reinforcement algorithms to improve the performance of cooperating agents in the L-ALLIANCE architecture by having the agents learn how to estimate the performance of other agents. Sen et al. [SSH94] use reinforcement learning in a two-robot box-pushing task. Balch [Bal97] applies reinforcement learning in building a task-achieving strategy for each robot in a team, while the robots learn individually to activate particular behavior assemblages given their current situation and a reward signal. Uchibe et al. [UAH98] propose a reinforcement learning method supported by system identification and learning schedules in multi-agent environments. Their method estimates the relationships between a robot's behavior and those of others through interactions. However, only one robot may learn and other robots will use a fixed policy in order for the learning to converge.

Reinforcement learning allows an autonomous agent that has no knowledge of a task or an environment to learn its behavior by progressively improving its performance based on given rewards as the learning task is performed. However, this may involve a high computational cost. Sometimes agents based on reinforcement learning can be very slow in adapting to environmental changes. Fukuda et al. [FFA99] propose a method to reduce the adaptation delay. Their idea is to make it possible for a robot to know if any environmental change has occurred through multiplex reinforcement learning. Sen et al. [SSH94] note that applying reinforcement learning to a group can make the learning more difficult as each individual agent becomes a dynamic factor in the others' learning. Takahashi and Asada [TA99] have developed a method of multi-layered reinforcement learning with which a control structure can be decomposed into smaller transportable chunks, and therefore previously learned knowledge can be applied to related tasks in newly encountered situations. Similarly, Fujii et al. [FAAE98] have applied a multi-layered reinforcement learning scheme to acquire collision-avoidance behaviors. Their scheme consists of four layers of modular control, corresponding to four stages of reinforcement learning that start with easier problems and proceed to more complicated ones.

It should be mentioned that the amount and quality of reinforcement can determine how quickly an agent will learn. In a nondeterministic, uncertain world, learning within bounded time requires reinforcement shaping in order to take advantage of as much information as available to the agent. As suggested by Mataric [Mat94a], there are generally two ways of accelerating reinforcement learning: (1) "by building in more information" and (2) "by providing more reinforcement." In the case of a multi-agent system, Yamaguchi et al. [YTY97] point out the importance of propagating and selectively sharing learning results. Along this line, Mataric [Mat97a] has studied how to learn social rules, including the observation and communication of a behavior, among four physical robots.

One of the attractive features of reinforcement learning lies in its formal foundation. If certain conditions (e.g., an infinite number of trials and a Markovian environment) are satisfied, an agent will converge toward an optimal action selection policy. Unfortunately, these conditions are seldom attainable in real, complex situations. As Maes [Mae95] points out, existing reinforcement learning algorithms will have several difficulties:

1. They cannot deal with time-varying goals; the action policy learned is for a fixed set of goals.
2. If the goals change, they will have to relearn everything from scratch.
3. For realistic applications, the size of the state space (or the number of situation-action pairs) is so large that learning takes too much time to be practical.
4. Only when a reward is received can the agent start learning about the sequence of actions leading to that reward; and as a result, it takes a lot of time to learn long action sequences.
5. Given faulty sensors or hidden states, it is difficult for the agent to know at all times which situation it is in.
6. It is hard to build initial knowledge into learning.
7. The agent cannot learn when multiple actions are taken in parallel.

Wyatt [Wya97] has also noted the limitations associated with reinforcement learning methods per se, and believes that these are especially pertinent to the problem of using them in real robots. First, while many interesting learning domains can be modeled as MDPs, agents situated in nondeterministic, uncertain environments do not fit in this model [Mat94a]. New algorithms have been designed to cope with this difficulty [LM92, WB90] but are not guaranteed to converge to an optimal policy under such circumstances. The second problem is that of slow convergence. Although deterministic MDPs can be solved efficiently, this has not been shown to extend to MDPs with stochastic transition functions [LDK95]. Furthermore, because the number of possible states increases exponentially with the number of features in the environment, the time taken to solve MDPs rises rapidly as the complexity of the environment increases. Consequently, in stochastic environments of large feature (and hence *state*) space, the convergence can be prohibitively slow. Two possible remedies to this problem will be: (1) to make the temporal credit assignment mechanism faster and (2) to make the temporal credit assignment problem simpler.

4.2 Genetic Algorithms

A genetic algorithm (GA) may be regarded as "a hill-climbing search method that finds near-optimal solutions by subjecting a population of points in a search space to a set of biologically inspired operators" [Gol89]. The basic principle of genetic algorithms works as follows [Hol75]:

1. The fitness of each member in a GA population is computed according to an evaluation function, called *fitness function*, that measures how well the individual performs in the given task domain.

2. The best individuals of the population are propagated proportionately to their fitnesses, while poorly performing individuals are reduced or eliminated completely.

3. By exchanging information between individuals to create new search points, the population explores the search space and converges to the neighborhood of an optimal solution.

The algorithm may find the optimal solution, but it is not guaranteed to do so. Genetic algorithms apply their operators to a representation of the search-space points. The representation is a position-dependent bit string, where each bit is a *gene* in the *chromosome* string. Several genetic operators have been proposed, but the three most frequently used are `reproduction`, `crossover`, and `mutation` [Gol89]. These operators are graphically illustrated in Figure 4.1. The reproduction operator "selects the fittest individuals and copies them exactly, replacing less-fit individuals so the population size remains constant." This increases the ratio of well-performing individuals to poorly performing ones. The crossover operator "allows two individuals to exchange information by swapping some part of their representations." The mutation operator is used to "prevent the loss of information that occurs as the population converges on the fittest individuals."

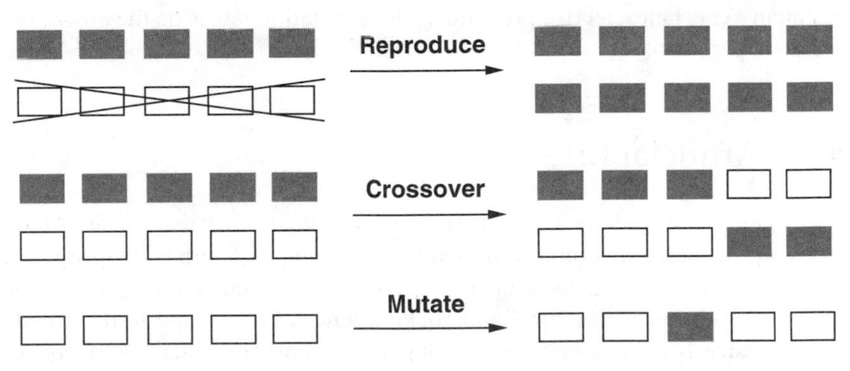

FIGURE 4.1. An illustration of genetic operators [RABP94].

Figure 4.2 presents the evolutionary cycle of GA in pseudo-code, where p_c and p_m specify the probabilities of crossover and mutation, respectively.

Generally speaking, every researcher has his or her own GA recipe. Some people rely solely on mutation and do not use crossover, whereas others stress the importance of crossover. Some prefer tournament-based selection, whereas others use global ranking of all individuals. Additionally, crossover can take different forms, depending very much on the genetic encoding scheme employed.

Ram et al. [RABP94] have explored the application of genetic algorithms in the learning of robot navigation behaviors. Unlike standard genetic algorithms, they

```
begin
  generation ← 0
  seed population
  while not termination condition do
    generation ← generation + 1
    calculate fitness
    selection
    crossover (p_c)
    mutation (p_m)
  endwhile
end
```

FIGURE 4.2. The evolutionary cycle in GA [Ang93].

use a floating point gene representation method. Dorigo and Schnepf [DS91] have used GA to train simulated robots to avoid obstacles and follow moving targets. Their genetic algorithm can determine when the robots should switch from one behavior to another.

Task domains do not always fall into the convenient picture of fixed-dimensional search space. In order to deal with such domains, SAGA (i.e., *Species Adaptation Genetic Algorithms*) [Har96] has been developed. It is an incremental evolution through a gradual increase in genotype length. Rank-based selection is used to maintain a constant selective pressure, and the mutation rate is on the order of one mutation per genotype.

4.3 Artificial Life

As originally defined by Langton [Lan88], the field of artificial life (AL) focuses on "the basic phenomena commonly associated with living organisms, such as self-replication, parasitism, evolution, competition, and cooperation." It complements traditional biological and social sciences by attempting to "simulate or synthesize lifelike behaviors in computers, robots, or other man-made systems." According to Langton, the goal of AL is to "model 'life-as-it-could-be' so as to enhance the understanding of 'life-as-we-know-it'." It increases our understanding of nature and offers us insight into artificial models, thereby providing us with the ability to improve their performance. More specifically, AL provides a unique framework for studying "how entities at different levels of organization (e.g., molecules, organs, organisms, and populations) interact among themselves" [Par97] although, of course, at the cost of introducing crude simplifications. AL draws on bottom-up modeling of complex systems, such as colonies of ant-like agents [BB99, DMC96]. It is this bottom-up, parallel, distributed, local determination of behavior that AL employs in its primary methodological approach to the generation of lifelike behaviors. The key concepts of AL are adaptation, self-organization, evolution, and emergent behaviors. Adaptation and self-organization mean that a system improves its own structure over time based on its own experience in an environment. Evolution offers the possibility of adaptation to the dynamic changes in the environment. When an unforeseen event occurs, the system can evolve and adapt to the new situation.

Although AL shares with AI its interest in synthesizing adaptive autonomous agents, the AL community has initiated a radically different approach toward the goal of building autonomous agents that focuses on fast, reactive behavior as well as adaptation and learning, rather than knowledge and reasoning [Lan88].

The AL approach differs from classical robotics approaches in that the former is interested in how a robotic agent interacts with its environment and learns from its interaction, leading to emergent robotic behaviors. Multi-agent robotics is related to AL in that both are concerned with exploiting the dynamics of local interactions between agents and the world in order to create complex global behaviors. However, research in AL usually deals with much larger population sizes.

Brooks [Bro92] has discussed the general issues involved in using AL techniques to program mobile robots. In particular, he examines the difficulties inherent in transferring programs evolved in a simulated environment to run on an actual robot. While analyzing the dual evolution of organism morphology and nervous systems in biology, he proposes a technique for capturing some of the search space pruning that the dual evolution offers in the domain of robot programming.

4.4 Artificial Immune System

Ishiguro et al. [IKW+97] point out that in behavior-based robotics, there are two problems that have to be resolved: how to construct an appropriate arbitration mechanism, and how to prepare appropriate competence modules. One of the promising approaches to tackle these two problems is based on a biological immune system – a biologically inspired behavior arbitration mechanism. This is because the immune system has various interesting features when viewed from the engineering standpoint, such as immunological memory, immunological tolerance, pattern recognition, and so on.

Recent studies on immunology have shown that "the immune system does not just detect and eliminate non-self materials called antigens (such as virus) but rather plays an important role in maintaining its own system against dynamically changing environments through the interactions among lymphocytes and/or antibodies" [Per92]. Therefore, the immune system can be expected to provide a new methodology suitable for dealing with unknown or hostile environments.

Following the descriptions in [AG96], the basic components of a biological immune system are macrophages, antibodies, and lymphocytes. Lymphocytes are mainly classified into two types, B-lymphocytes and T-lymphocytes. B-lymphocytes are the *robots* maturing in bone marrow.

> Roughly 10^7 distinct types of B-lymphocytes are contained in a human body, each of which has a distinct molecular structure and produces Y-shaped antibodies from its surfaces. The antibody recognizes specific antigens, which are the foreign substances that invade a living system. To cope with continuously changing environments, living systems possess an enormous repertoire of antibodies in advance. On the other hand, T-lymphocytes are the *robots* maturing in thymus, and they generally perform to kill infected *robots* and regulate the production of antibodies from B-lymphocytes.

Immune system models can be extended to multi-agent robotic systems. Mitsumoto et al. [MFA+96] have drawn parallels at several levels – a robot and its environment are modeled as a stimulating antibody-antigen relationship, and robot-robot interactions can be both stimulating and suppressing (analogous to antibody-antibody relationships). Each robot decides its next action based on its relationships with other robots and the world, organizing itself to effectively conduct a task. Fukuda et al. [FMSA99] have utilized an immune architecture to realize a phase change of group behavior patterns in Micro Autonomous Robotic System (MARS) and design lifelike robotic systems.

4.5 Probabilistic Modeling

The core of probabilistic modeling is the idea of representing information through probability densities. Specifically, Thrun [Thr00] emphasizes two aspects of probabilistic characterization in robotics:

1. **Probabilistic perception:** Robots are inherently uncertain about the state of their environments. The uncertainty arises from sensor limitations, noise, and the fact that most interesting environments are, to a certain degree, unpredictable. When *guessing* a quantity from sensory data, probabilistic approaches compute a probability distribution over what might be the case in the world instead of generating a single *best guess* only. As a result, a probabilistic robot can gracefully recover from errors, handle ambiguities, and integrate sensory data in a consistent way. Moreover, the probabilistic robot knows about its own ignorance – a key prerequisite of autonomous robots.
2. **Probabilistic control:** Autonomous robots must act in the face of uncertainty. Instead of considering the most likely situations only (current or projected), probabilistic approaches strive to compute a decision-theoretic optimum, in which decisions are based on all possible contingencies.

Probabilistic approaches are typically more robust in the face of sensor limitations, sensor noise, and environment dynamics [BIM00]. They often scale better to complex environments, where the ability to handle uncertainty is of greater importance. In addition, "probabilistic algorithms make much weaker requirements on the accuracy of models than many classical planning algorithms do, thereby relieving a programmer from the (unsurmountable) burden of coming up with accurate models" [BIM00]. As viewed probabilistically, the robot learning is a long-term estimation problem. Thus, probabilistic algorithms provide a sound methodology for many flavors of robot learning. However, probabilistic algorithms are inherently less efficient than non-probabilistic ones due to the fact that they consider entire probability densities.

Probabilistic approaches are similar to behavior-based approaches since they place a strong emphasis on sensory feedback [BIM00]. Because probabilistic models are insufficient to predict an actual state, sensor measurements play a vital role in state estimation and, thus, in determining a robot's actual behavior. At the same time, they differ from behavior-based approaches since they rely on planning and since the robot behavior is not just a function of a small number of recent sensory readings.

Burgard et al. [BMF+00] have presented a probabilistic approach for the coordination of multiple robots that simultaneously explore an unknown environment. They take into account the costs of reaching a target point and the utility of the target point, as given by the size of the unexplored area that a robot can cover with its sensors upon reaching the target position. Whenever a target point is assigned to a specific robot, the utility of the unexplored area visible from this target position is reduced for other robots. In this way, a team of robots assigns different target points to different individuals.

Simmons et al. [SAB+00] address the problem of exploration and mapping of an unknown environment by multiple robots with a mapping algorithm and an exploration algorithm. The former is "an online likelihood maximization that uses hill climbing to find maps that are maximally consistent with sensory data." The latter explicitly coordinates the robots. It tries to maximize the overall utility by minimizing the potential of overlap in information gain among the various robots. Thrun et al. [TFB98] consider map building as a "constrained, probabilistic maximum-likelihood estimation problem" and address the task of building large-scale geometric maps of indoor environments with mobile robots. They have proposed a practical algorithm for generating the most likely map from data, along with the most likely path to be taken by a robot.

Localization is a process of updating the pose of a robot in an environment based on sensory readings. Fox et al. [FBKT99, FBKT00] present a probabilistic algorithm for collaborative mobile robot localization. They use a sample-based version of Markov localization, capable of localizing mobile robots in an anytime fashion. When a team of robots localizes in the same environment, probabilistic methods are employed to synchronize each robot's belief whenever one robot detects another. As a result, the robots can localize themselves faster.

4.6 Related Work on Multi-Robot Planning and Coordination

Besides the above-mentioned work, some well-known strategies, such as multi-agent planning and model-based techniques, can also play an important role in multi-agent robotic systems. Inoue et al. [IOH98] have described a planning method for an interactive transportation task by cooperative mobile robots in an unknown environment. The task requires the acquisition of environmental information, the generation of appropriate robot paths based on the acquired information, and the formation of a robot group. In order to realize an efficient transportation, they propose a motion planning architecture consisting of *environmental exploration phase*, *path-generation phase*, and *strategy-making phase*. In each phase, every robot plans its own motion individually; and the phase transitions are made at the same time for all robots. Miyata et al. [MOA$^+$00] have developed a task-assignment architecture for dealing with the same task. They consider three needs that a planner should meet: to deal with a variety of tasks in time and space; to deal with a large number of tasks; and to decide behavior in real time. The following approaches have been proposed:

1. Based on sensory information, tasks will be dynamically generated using so-called task templates.
2. The generation of tasks is tuned in quality by feeding back executed results.
3. The main part of the architecture consists of two real-time planners: a priority-based task-assignment planner using a linear programming method and a motion planner based on short-time estimation.

Donald et al. [DJR93] have examined motion-planning algorithms for coordinated manipulation with different numbers of robots and different amounts of *a priori* knowledge about an object to be moved. The theoretical aspect of their work focuses on computing information requirements for performing particular robot tasks such as box-pushing. Botelho and Alami [BA00] have proposed a general architecture called $M+$ *cooperative task achievement*, where various schemes for multi-robot task achievement are integrated. It is based on an online combination of local individual planning and multi-robot plan validation for coordinated and cooperative behaviors. The robots plan/refine their respective missions, taking into account other robots' plans and social rules as planning/refinement constraints, and thus produce validated multi-robot plans containing coordinated and cooperative actions.

Last but not least, Goldberg and Mataric [GM99] have shown how various levels of coordinated behaviors may be achieved in a group of mobile robots by applying a dynamics model of the interaction between a robot and its environment. They use augmented Markov models (AMMs) as a tool for capturing such interaction dynamics online and in real time, with little computational and storage overhead.

5

Outstanding Issues

There are many other deep, simple principles: continuity, connectivity, feedback, information, order, disorder, bifurcation, learning, autonomy, emergence....

Right now, we are aware of a few of these deep principles. We need more. We also need a better understanding of how to use those principles. We also need to extend the range of systems that our mathematics can handle.[1]

Ian Stewart

[1] *Life's Other Secret*, John Wiley & Sons, 1998, p 247.

5.1 Self-Organization

One of the main features that makes the multi-agent approach attractive in synthesizing the adaptive behavior of a multi-robot system is the possibility of self-organization [Nol98]. The advantages of self-organization and the efficiency of self-organizing behaviors in animal societies can be readily noted. Some animal societies, particularly social insects, can achieve complex tasks that are impossible to complete individually [BDT99]. In a self-organizing system, simple programs can operate in unforeseen situations and adapt to changing conditions.

Bonabeau et al. [BDT99] summarized three important characteristics of self-organization.

> First, a self-organizing system can accomplish complex tasks with simple individual behaviors. Second, any change in the environment may influence the same system to generate a different task, without any change in the behavior characteristics. Finally, any small differences in the individual behaviors can influence the collective behavior of the system.

Therefore, social complexity of the system is compatible with simple identical individuals, as long as the communication among the individuals can provide a necessary amplifying mechanism. For example, a swarm of robotic agents gathering under a palletized load can change the operation *phase* by receiving a signal from any member of the swarm. Bonabeau et al. [BTC98, BT95] achieved this by defining specific communication mechanisms. Generally speaking, the characteristics of self-organization are desirable in a swarm of robots where simple individual behaviors can be achieved with relatively low cost [Bon98]. At the same time, the simplicity (and homogeneity) of individual agents in a robotic swarm also decreases the likelihood of breakdown. Also, the breakdown of one agent will not affect the activity of the whole robotic swarm [Ste95].

5.2 Local vs. Global Performance

With respect to multi-robot performance, several important issues can be posed for a system of decentralized autonomous robots.

Given a finite number of behaviors locally defined for individual robots in an environment, how will the robots converge to a set of desirable configurations? How will the parameters, such as the initial number/distribution of the robots and their given behavior parameters, affect the converged states?

Suppose that we have two concurrent ways of linearly changing behavior parameters, in order to make the steady state convergence faster and more selective (since we may be interested in only one of the states). The first way is through each of the individual robots – the robot records its own performance, such as the number of encounters and the number of moves. Then, based on such observations, it tries to modify its own behavior parameters in order to achieve an optimal performance, such as the maximum number of encounters and the minimum number of moves. Another way is through the feedback of the information that is observed globally from the whole system, such as the spatial pattern formation in group robots. For example, a particular group of robots switches from one behavior to another as commanded by a global control mechanism; or the behavior parameters in a particular group change in a certain way. From the above descriptions, we now come to the following questions: in order to achieve an optimal multi-robot performance, how much optimization at the local individual level and how much at the global level will be necessary? What will be a reasonable balance between the two? And, how do we dynamically maintain such a balance?

5.3 Planning

The bottom-up organization of robot behaviors has several variations. For example, *supervenience architecture* adds a goal-driven mechanism to enable a hierarchical task-network planner to adapt behaviors to a context and to new goals [SH94]. Similarly, Arkin's AuRA binds a set of reactive behaviors to a simple hierarchical planner that chooses appropriate behaviors in a given situation [Ark89]. Toward an integration of planning and reacting, Lyons and Hendriks [LH94] have developed a system called *RS* in which planning is seen as the permanent adaptation of a reactive process.

To date, research in multi-agent planning has been limited primarily to the areas of distributed artificial intelligence (DAI) [SV97, SV98, SL93, DMC96]. With DAI, several agents may cooperate to accomplish a certain task. The task may be one of such complexity that no single agent can accomplish it alone. Alami et al. [AFH+98] examined the planning and plan cooperation in multi-agent robotic systems and have identified the following issues:

1. **Global vs. local**: In planning actions for a group of robots, we can consider the whole group or limit the scope of planning to the subgroups of robots with conflicting actions. If the number of critical resources in an environment is more or less equal to the number of robots, conflict resolution may, by propagation, involve the whole group. On the other hand, if the environment is properly sized, conflicts may remain local; and the solutions can be negotiated locally without disturbing unconcerned robots. In addition, the uncertainty in perceiving states will grow with the increased complexity of the environment. Consequently, global planner-based approaches to control will not be well suited for problems involving multiple robots acting in real time based on uncertain sensory information.
2. **Complete vs. incremental**: We can also limit the scope of planning and plan cooperation in time. When a mission is sent to a robot, the robot may plan the whole mission. But considering the execution

hazards and computational costs involved, it will be inefficient to plan too far ahead. Plan cooperation can be done continuously to guarantee a smooth navigation and to avoid over-constraining other robot plans.

3. **Centralized vs. distributed**: Where should planning and plan cooperation take place, in a centralized computer or on board? This does not change the computational complexity of treatment itself. However, in a centralized approach, all the data (which is mostly local) needs to be sent to a central station and therefore requires a reliable communication link with a high bandwidth between robots and the central station.

5.4 Multi-Robot Learning

While a considerable amount of work has been devoted to multi-agent learning [Wei96], somewhat less work has been accomplished to date in multi-robot learning. Multi-robot learning is a complex problem. As Schaal [Sch00] has noted, research needs to address how to learn from (possibly delayed) rewards, how to deal with high-dimensional learning, how to use efficient function approximators, and how to embed all the elements in a distributed control system with real-time performance. A particularly challenging domain for multi-robot learning concerns the cooperation of team robots.

5.5 Coevolution

Coevolution has been receiving increased attention as a method of multi-agent simultaneous learning (e.g., to evolve a learner and a learning environment simultaneously). Mataric [Mat94b] has applied coevolution in a group of physical robots to achieve collective learning through a direct exchange of received reinforcement and learned information.

As highlighted by Nolfi and Floreano [NF98], coevolution has several interesting features that can potentially enhance the adaptation power of artificial evolution. For instance, the performance of individuals in a population depends on the individual strategies of other populations, which vary during an evolutionary process.

Existing coevolution methods have mostly focused on two competing individuals, such as a prey and a predator [CM96, FN97, LHF+97]. Since behaviors emerged from a multi-agent environment include not only competition but also cooperation, ignorance, and so on, artificial coevolution beyond competition should also be explored. Uchibe et al. [UNA98] suggest that the environment itself should coevolve from simpler to more complicated situations and thus assist agents in developing the desired skills of competition and cooperation. More systematic studies are needed to understand what are the necessary and sufficient conditions for leading coevolutionary processes to successful results.

5.6 Emergent Behavior

Steels [Ste95] has noted two advantages of emergent behavior when compared to directly programmed behavior:

1. No additional structure is needed inside an agent to get additional capabilities. Therefore, we do not need any special explanations on how the behavior may come about.
2. Emergent behavior tends to be more robust because it is less dependent on accurate sensing or action and because it makes less environmental assumptions.

Behavior-based AI research considers the notion of emergent behavior as a possible explanation for the emergence of functional complexity in agents. Emergent behavior implies a holistic capability where "the sum is greater than its parts." Arkin [Ark98] provided a survey of discussions on emergence:

1. Emergence is "the appearance of novel properties in whole systems" [Mor88].
2. "Global functionality emerges from the parallel interaction of local behaviors" [Ste90].
3. "Intelligence emerges from the interaction of the components of the system" (where the system's functionality – planning, perception, mobility, etc. – results from the behavior-generating components) [Bro91].
4. "Emergent functionality arises by virtue of interaction between components not themselves designed with the particular function in mind" [MB93].

Furthermore, Arkin [Ark98] defines emergence as "a property from a collection of interacting components – behaviors." Generally speaking, emergent properties are common phenomena in behavior-based systems. In some cases, they are straightforward, such as choosing the highest ranked or most dominant behavior; in others they are more complex, involving a fusion of multiple active behaviors. Further studies are needed in order to better understand the efficient ways of inducing emergent properties in a multi-robot system.

5.7 Reactive vs. Symbolic Systems

Spier [Spi97] categorized researchers in cooperative robotics into those who believe "robot control is best achieved through symbolic means – including explicit world representation and logical reasoning" and those who believe "it is best achieved through reactive means – in which robots rely on simple behaviors and intelligence emerges naturally from the interactions among those behaviors." As he points out, both approaches have their drawbacks.

> While reactive systems are robust and understandable (at the single-agent level), they are generally inefficient and extremely complex (at the global level) – especially when a global behavior is desired, as in many multi-agent systems.

Often, reactive systems seem to attain a correct global behavior through a combination of luck and sheer persistence on the part of a programmer. As a reactive rule base grows, reasoning about the complex interactions among the rules becomes very difficult [OJ96].

On the other hand, symbolic systems, which generally perform more predictably than reactive systems, and whose global behavior is easier to understand than that of an equivalent reactive system, also have a number of problems. As Spier [Spi97] indicates, they are not robust; the failure of one component in a multi-agent planning system can lead to the failure of the whole system. In addition, difficult tasks can lead to poor performance on the part of symbolic systems.

Generally speaking, reactive approaches [LVCS93, Mat95a, Mat95b, SB93, BDT99] tend to view cooperating agents as decentralized groups of peers. In these approaches, each agent follows its own reactive programming. Besides robustness, reactive systems are desirable in that they are modular (theoretically, the programmer needs only to think in terms of a single robot) [Mat92a, Mat95b, SB93].

5.8 Heterogeneous vs. Homogeneous Systems

Heterogeneity in multi-robot systems presents a challenge to efficient autonomous control, especially when an overlap in team member capabilities exists. As Mataric [Mat94a] observes, in a multi-robot system, more than one robot may be able to perform a given task, but "with different levels of efficiency." In such a case, the robots must continuously determine which individual in the team is currently best suited for the task. This type of decision is usually not so easy to make, especially when the multi-robot system control is distributed across all team members. Thus,

> A heterogeneous team control mechanism must have some effective means of distributing tasks so that an acceptable level of efficiency can be achieved without sacrificing the desirable features of fault tolerance and robustness [Mat94a].

Most research in multi-robot systems has centered on homogeneous systems, with some work on heterogeneous systems focusing on mechanical and sensor differences (for example, see Parker's work [Par94]). Recent investigations indicate that behaviorally heterogeneous systems can offer advantages in certain tasks [Bal97, FM97]. Teams of mechanically identical robots are particularly interesting because they can be either homogeneous or heterogeneous, depending only on their behavior.

Goldberg and Mataric [GM99] proposed a framework for investigating the relative merits of heterogeneous and homogeneous behaviors in foraging tasks. They focus on mechanically identical but behaviorally different agents. Time, interference, and robustness are proposed as *metrics* for evaluating a foraging robot team, while pack, caste, and territorial arbitration are offered as mechanisms for
generating an efficient behavior.

5.9 Simulated vs. Physical Robots

There are good reasons for working with simulated robots: "learning often requires experimenting with behaviors that may occasionally produce unacceptable results" [Sch94]. Making mistakes on real physical systems can be quite costly and dangerous. In addition, the evolution of complex behaviors from scratch on a physical robot, even if it is technically feasible, would require "an adaptation time that is too long to be practically exploited for real-world applications" [FM98]. It is well-known that the performance of genetic algorithms is sensitive to initial conditions. Evolution in simulated environments, usually less time-consuming than that in the real world, may allow us to ascertain to what extent a specific evolutionary process is sensitive to the initial conditions and therefore the probability that a limited number of simulations in real environments can produce desired performance. Nolfi et al. [NFMM94] suggest that simulated robots be used in developing control systems for real robots when certain special conditions are taken into account. However, one should not expect control systems that are evolved in a simulated environment to behave exactly the same in a real environment.

As researchers have pointed out [Bro92, FM96], there are certain situations where those who use simulated robots to develop control systems for real robots may encounter problems:

1. Sensors should not be confused, as in simulation. However, physical sensors do not separate objects from the background, they do not operate in a stable coordinate system, and they do not provide information regarding the absolute positions of objects.
2. Simulation usually does not consider all the physical laws of the interaction of a real agent with its environment, such as mass, weight, friction, inertia, etc.
3. Physical sensors deliver uncertain values and commands to actuators that may have very uncertain effects, whereas simulation often uses grid-worlds and sensors that return perfect information.
4. Different physical sensors and actuators perform differently because of the slight differences in electronics and mechanics or because of their different positions in a robot. This factor is not considered in simulation.

Generally speaking, simulation can be of great help when properly integrated with tests on physical robots. Current simulation methods have progressed from unrealistic grid-worlds to new methodologies that, to a certain extent, guarantee an acceptable transfer to a target robot under well-understood constraints [Jak98a, Jak98b, NFMM94].

5.10 Dynamics of Multi-Agent Robotic Systems

Cooperative behaviors in multi-agent robotic systems arise out of the local actions and interactions of relatively simple (as compared to the system as a whole) individuals, without the existence of centralized control [Win90, NNS97]. Often, the local interactions are nonlinear. In other words, the behavior of the system as a whole is more than a linear superposition of the behaviors of the individuals when considered in isolation. This property can be stated using the popular phrase "the whole is more than the sum of the parts" [Hor99]. The global, cooperative behavior-generating dynamics in multi-agent robotic systems may be regarded as an emergent property of such systems.

Harvey [Har00] defines a dynamic system as "any system with a finite number of state variables that can change over time." The rate of change in one variable depends on the current values of some or all of the variables in a regular fashion. The design problem that a robot builder faces is creating the internal dynamics of a robot as well as the dynamics of its coupling and its sensorimotor interactions with its environment, such that the robot can exhibit a desired behavior. The design of any one component of a control system depends on an understanding of how it interacts in real time with other components. Such an interaction may be mediated via the environment [Bro99]. Brooks' subsumption approach [Bro86] offers one design heuristic:

> First build a robot with behaviors simple enough to understand, and then incrementally add new behaviors of increasing complexity or variety, one at a time, which subsume the previous ones. Before the designer adds a new control system component in an attempt to generate a new behavior, the robot is fully tested and debugged for its earlier behaviors.

Huberman and Hogg [HH93] studied a form of distributed computation in which agents have incomplete knowledge and imperfect information about the state of the whole system, and they presented a mechanism for achieving global stability through local control. They [HH95] also proposed a detailed model of collaboration in communities and its dynamics. This model enables a community to naturally adapt to its growth and specialization or to any changes in the environment without the need for any centralized control. Youssefmir and Huberman [YH95] have shown the dynamics of a multi-agent system whose members continuously modify their behaviors in response to the dynamics that unfold within the system. The behaviors are driven by the local optimization of the utilities that the agents accrue when gaining access to resources. The agents decide on the basis of bounded rationality, which implies imperfect information and delayed knowledge about the present state of the system.

5.11 Summary

Multi-agent robotic system design is challenging because the performance of such a system depends significantly on issues that arise from the interactions between robots. Distributed approaches are appealing due to their properties of scalability and reliability. Learning, evolution, and adaptation are three fundamental characteristics of individuals in a multi-agent robotic system. Behavior-based robotics, collective robotics, and evolutionary robotics have offered useful models and approaches for cooperative robot control in the multi-robot domain. They are inspired by the ideas from biology, ethology, and sociology. Different techniques of reinforcement learning, genetic algorithms, artificial life, immune systems, probabilistic approaches, and multi-agent planning can be helpful in the design of cooperative controllers for multiple robots. Several important issues in multi-agent robotics, such as self-organization, multi-agent planning and control, coevolution, emergent behavior, reactive behavior, heterogeneous/homogeneous design, multi-agent simulation, and behavioral dynamics need to be addressed. The ultimate aim is to synthesize complex group behaviors from simple social interactions among individuals based on simple strategies.

Part II

Case Studies in Learning

6

Multi-Agent Reinforcement Learning: Technique

Learning itself may be the fundamental mechanism which converts chaotic attractors to orderly ones.[1]

Stuart A. Kauffman

What is a robot group? How do we characterize group robots? What is the mechanism for collective learning? How do we design such a mechanism? This chapter will address these questions. The models and mechanisms introduced here will serve as a basis for the following discussions in Chapters 7 and 8. Specifically, we will consider the building blocks of group robots, consisting of sensing capability, primitive behaviors, behavior learning mechanism, and behavior selection mechanism.

[1] *The Origins of Order*, Oxford University Press, Inc., Oxford, 1993, p 229.

6.1 Autonomous Group Robots

6.1.1 Overview

Figures 6.1 and 6.2 provide two schematic diagrams showing a group of autonomous robots and the architecture of an individual robot, respectively.

As illustrated in Figure 6.2, each robot carries a ring of equally spaced sensors that enables the robot to get the proximity information from its local environment. The extracted proximity information is called external *stimulus* to the robot. In addition to sensors, the robot has two driving wheels and one omnidirectional wheel that allow the robot to move in any arbitrary direction with an arbitrary step once it is commanded to do so. The robot also has an on-board behavior selector. The functions of the behavior selector will be discussed in the following sections.

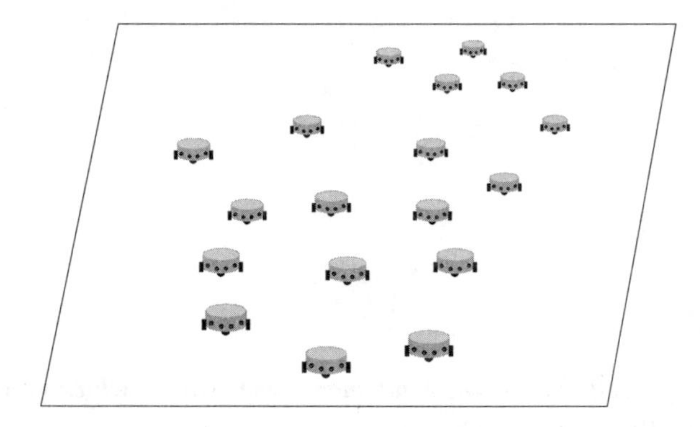

FIGURE 6.1. A group of autonomous robots in an environment.

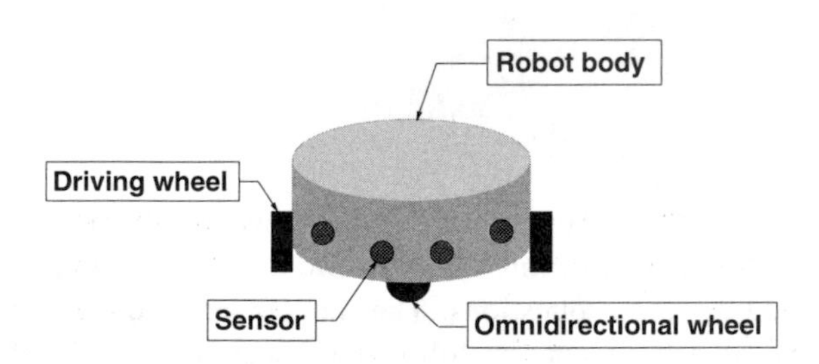

FIGURE 6.2. A schematic diagram illustrating a mobile robot (©1998 IEEE).

6.1.2 *Sensing Capability*

Each robot sensor can detect whether other robots are present in a sector as measured by angle β within a certain distance. The surrounding area within which the sensor can detect other robots is called a neighboring region. When there is/are robot(s) inside this region, the sensor will return 1; otherwise, it will return 0. In order to be able to instantaneously determine the presence or absence of other robots in a local two-dimensional environment, the robot utilizes all $\mathcal{N}$ equally spaced sensors, where

$$\mathcal{N} = \frac{2\pi}{\beta}. \tag{6.1}$$

The proximity information acquired from the $\mathcal{N}$ sensors, with respect to a specific group of robots, will be represented into a binary string of length $\mathcal{N}$. The most significant bit (MSB) in the binary string corresponds to the first sector in a polar coordinate system of the robot. Figure 6.3 illustrates an example where $\mathcal{N}$=8.

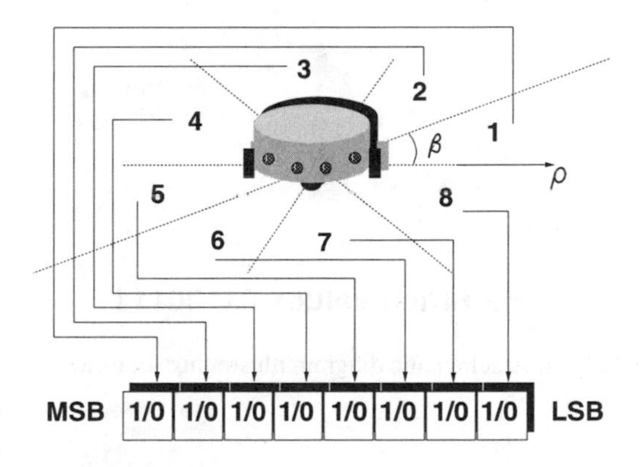

FIGURE 6.3. Eight proximity sensors mounted on a robot and their corresponding representation as a binary string.

Each group robot can choose either long- or short-range sensing capabilities. The difference is that the long-range capability can sense a large region comparable to the size of the entire task environment, whereas the short-range capability can extract proximity information only within a limited range of radius R_0.

6.1.3 *Long-Range Sensors*

A long-range sensor is a sensor that acquires the information on whether other robots are present inside a circular sector of a very large radius (which may be regarded as infinite). For this type of sensor, the returned value of sensor j in sector $\mathcal{E}_j$ is expressed as follows:

$$r_j = \begin{cases} 1, & \exists P_i \in \mathcal{E}_j, \\ 0, & \forall P_i \notin \mathcal{E}_j, \end{cases} \tag{6.2}$$

where P_i denotes the position of robot i. An arrangement of $r_j, \forall j = 1, 2, ..., \mathcal{N}$, gives a binary string of length $\mathcal{N}$. All possible binary strings provide the complete set $\mathcal{R}$ of returned values, that is:

$$[r_1, r_2, \cdots, r_{\mathcal{N}}] \in \mathcal{R}. \tag{6.3}$$

Figure 6.4 presents an example of sensing results in an environment, where each robot has eight long-range sensors.

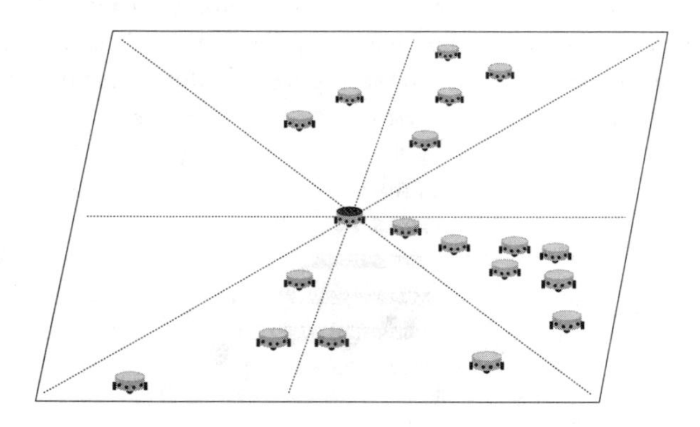

Returned value: 01100111

FIGURE 6.4. A schematic diagram illustrating long-range sensors.

6.1.4 Short-Range Sensors

A short-range sensor is a sensor that acquires the information on whether other robots are present inside a surrounding region of a certain radius, R_0. In this case, a circular sector j in the surrounding region of a sensing robot is denoted as $\breve{\mathcal{E}}_j$ and expressed as follows:

$$\breve{\mathcal{E}}_j = \{P_i \mid \; | P_i - P_0 | < R_0\}, \tag{6.4}$$

where P_i denotes the position of robot i, and P_0 denotes the position of a sensing robot.

The returned value of sensor j in sector $\breve{\mathcal{E}}_j$ can be expressed as follows:

$$r_j = \begin{cases} 1, & \exists P_i \in \breve{\mathcal{E}}_j, \\ 0, & \forall P_i \notin \breve{\mathcal{E}}_j. \end{cases} \tag{6.5}$$

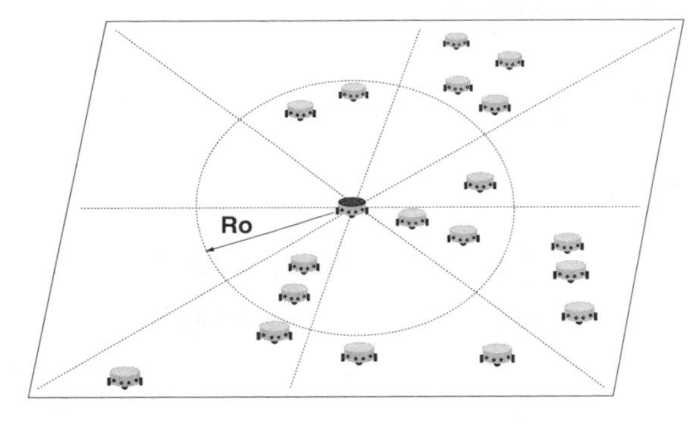

Returned value: 10100101

FIGURE 6.5. A schematic diagram illustrating short-range sensors.

Figure 6.5 presents an example of sensing results in an environment, where each robot has eight short-range sensors.

It should be mentioned that short-range sensors can be further divided into two categories:

1. In the first category, the reachable distance is fixed as a constant, R_0, which is equal to an original sensing radius.

2. In the second category, the reachable distance is adjustable. Such a sensor searches the surrounding region of a robot with an original sensing radius first. If all the sensors return 0, then the sensing radius of these sensors will increase by an increment, ΔR, based on the previous sensing radius. The robot will adjust its sensing range in this way until the returned value of some sensor(s) is 1. Figure 6.6 presents an operational flowchart for this category of sensors.

6.1.5 Stimulus Extraction

Each returned binary string of length $\mathcal{N}$ corresponds to a local condition of the robot environment. Although there can be as many as $2^{\mathcal{N}}$ possible string values, the number of distinct local conditions that can actually occur will be quite limited. For instance, if we perform *circular shift* (to left or to right) or *circular reverse shift* operations as illustrated in Figure 6.7, we realize that for each binary string (except strings 00000000 and 11111111), there exists an equivalent string. Two strings are considered equivalent if their expressed proximity distributions surrounding a sensing robot are the same. Figure 6.9 shows two equivalent strings: 00000011 and 00011000. A group of equivalent binary strings obtained by performing the `shift` operations corresponds to a distinct local condition, called a *stimulus*.

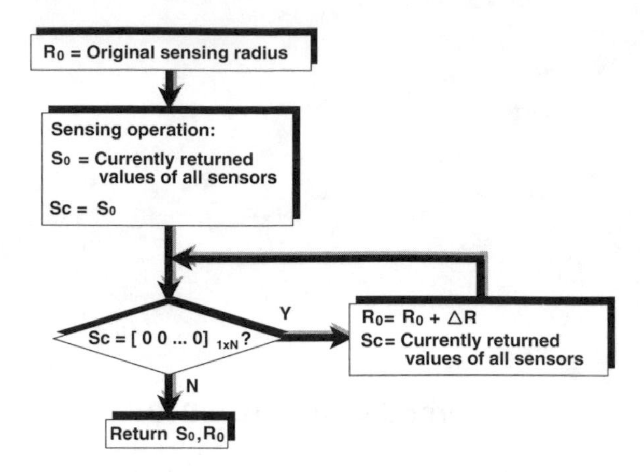

FIGURE 6.6. An operational flowchart for a radius-adjustable short-range sensor.

We define an operator, `extr` (i.e., extraction), to map from the set of returned sensory measurements $\mathcal{R}$ to the set of stimuli $\mathcal{S}$:

$$\texttt{extr} : \mathcal{R} \rightarrow \mathcal{S}. \tag{6.6}$$

Table 6.1 gives the number of possible stimuli with respect to string length $\mathcal{N}$.

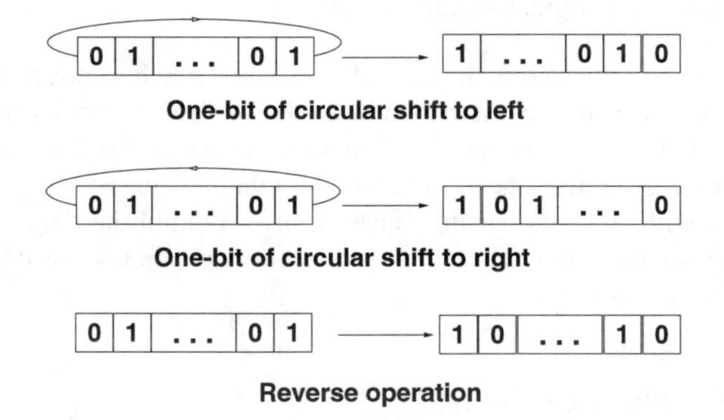

FIGURE 6.7. A schematic diagram illustrating circular `shift` and `reverse` operations. The corresponding algorithms are given in Figure 6.8.

6.1.6 *Primitive Behaviors*

A group robot can move in an arbitrary direction with an arbitrary step. However, in order to conveniently control the robot, we assume that the motion direction of the robot is divided into $\mathcal{N}$ sectors, corresponding to $\mathcal{N}$ equally spaced sensors. Figure 6.10 shows the motion directions when $\mathcal{N}$=8, with respect to a local polar coordinate system. The motions in these directions are referred to as the *primitive*

```
# Circular Shift to Left for M (0<M<N) bits:

   NewCodeString(1:(N-M))=OldCodeString((M+1):N);
   NewCodeString((N-M+1):N)=OldCodeString(1:M);

# Circular Shift to Right for M (0<M<N) bits:

   NewCodeString(1:M)=OldCodeString((N-M+1):N);
   NewCodeString((M+1):N)=OldCodeString(1:(N-M));

# Reverse Operation:

   for i=1:N
     NewCodeString(i)=OldCodeString(N-i+1);
   end
```

FIGURE 6.8. The algorithms for circular `shift` and `reverse` operations.

$\mathcal{N}$	1	2	3	4	5	6	7	8	9	...
Total	2	3	4	6	8	14	18	29	42	...

TABLE 6.1. The number of possible stimuli with respect to string length $\mathcal{N}$.

behaviors of the robot. We will use an $\mathcal{N}$-dimensional vector to represent the primitive behaviors of the robot:

$$[Behavior] = \begin{bmatrix} B_1 \\ B_2 \\ \vdots \\ B_{\mathcal{N}} \end{bmatrix}, \qquad \text{where } B_i \in \{-1, 1\}. \tag{6.7}$$

If a component of this vector, $B_i = 1$, it means the robot is capable of performing behavior i. On the other hand, $B_i = -1$ indicates that the robot cannot generate behavior i.

Corresponding to an array of robot primitive behaviors, we will further define a weight vector to express the probability of success if a specific behavior response is performed given a certain stimulus. The weight vector can be written as follows:

$$W = \begin{bmatrix} w_1 \\ w_2 \\ \vdots \\ w_{\mathcal{N}} \end{bmatrix}, \tag{6.8}$$

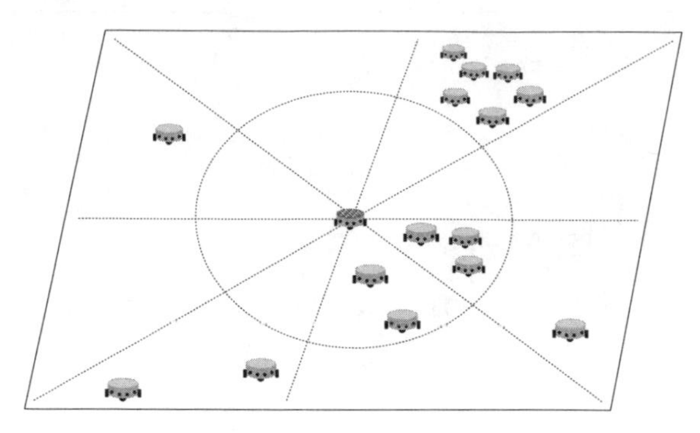

(a)

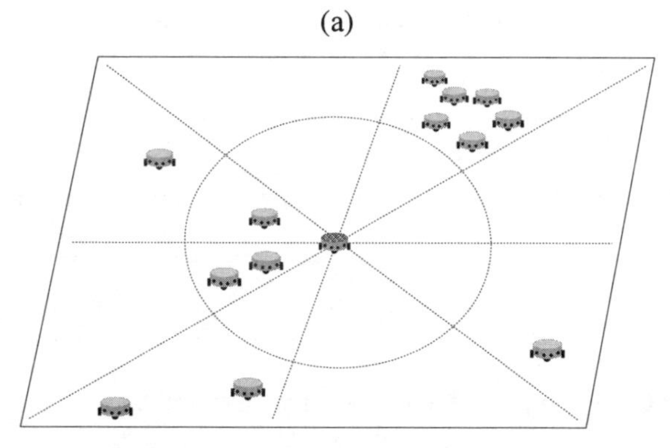

(b)

FIGURE 6.9. Two equivalent binary strings and their corresponding local conditions.

where

$$\begin{array}{ll} w_i = -1, \quad \text{if } B_i = -1, \quad i = 1, 2, \cdots, \mathcal{N}, \\ \sum_{i=1}^{\mathcal{N}} w_i \mid_{w_i \neq -1} = 1. \end{array} \tag{6.9}$$

It should be mentioned that the actual movement step size d_0 of each robot at a given time will depend on the maximum movement step size d_m and the maximum free movement step size d_p, as follows:

$$d_0 = \mathtt{min}(d_m,\ d_p), \tag{6.10}$$

where d_m is determined according to a predefined ratio η, i.e.,

$$\eta = \frac{d_m}{R_0}, \tag{6.11}$$

and the maximum free movement step size d_p is determined by the maximum collision-free displacement in a chosen direction. Figure 6.11 illustrates the

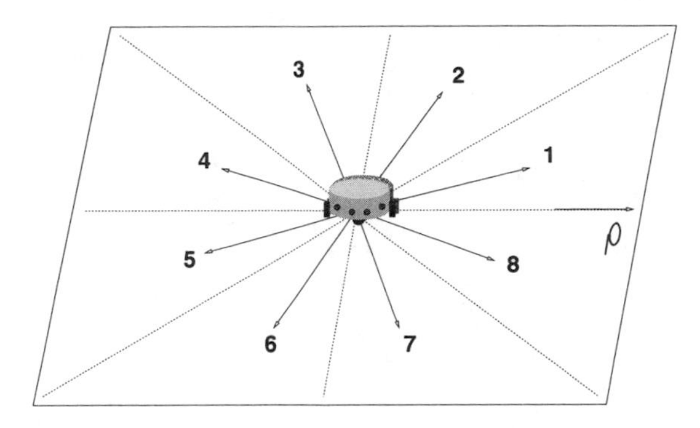

FIGURE 6.10. A schematic diagram illustrating the primitive behaviors of a robot.

execution of a behavior (i.e., a movement in a certain direction), where P^t and P^{t+1} denote the positions at time t and time $t + 1$, respectively.

6.1.7 Motion Mechanism

Having defined the notion of primitive behaviors and its associated representations, let us now consider some possible motion mechanisms, as follows:

1. **Experience-driven motion**: A robot with this control mechanism moves in its local environment without any predefined conditions. The robot selects the best primitive behavior based on its current stimulus and previously acquired experience. It executes the selected behavior and then updates its experience based on an evaluation of the behavior.

2. **Conditional motion**: A robot with this control mechanism selects and executes a certain behavior whenever some predefined conditions are satisfied. These conditions are the policies according to some special stimuli. After the execution of the selected behavior, it learns some aspects of this behavior based on a behavior learning mechanism.

3. **Random motion**: A robot with this control mechanism moves in a random manner. There is no learning or experience updating involved in this mechanism.

4. **Stationary**: A robot with this control mechanism does not execute any motion. Its kinetic energy is equal to zero.

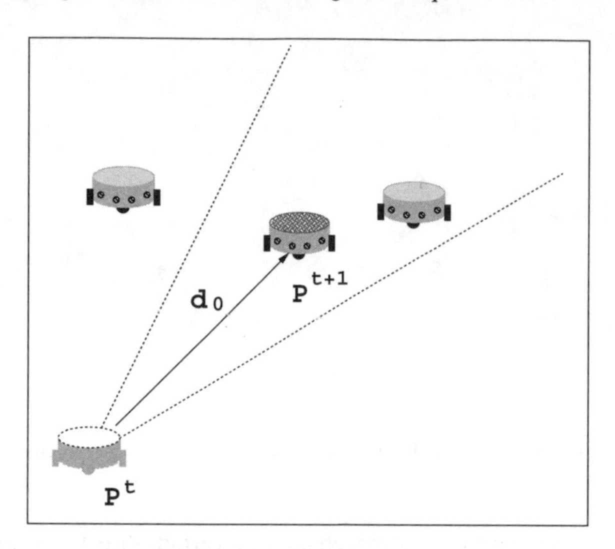

FIGURE 6.11. A schematic diagram illustrating behavior execution (©1998 IEEE).

6.2 Multi-Agent Reinforcement Learning

The sensory data received by an autonomous mobile robot from its environment constitutes a stimulus (or behavior triggering condition) to the robot. Based on its embedded learning mechanism, the robot will derive a specific stimulus-response behavioral association through behavior weight updating, as illustrated in Figure 6.12. Given multiple triggering conditions, the robot will determine its next move based on the result of voting from multiple behavioral associations, as illustrated in Figure 6.13.

Any response acquired in a situation will be represented in a vector of varying weights, called a behavior weight vector. In other words, given a stimulus, $\int_i$, there exists a vector, W, where each component expresses the probability of having a good performance if a certain reactive motion is executed. This pair of $\int_i$ and W is referred to as a *stimulus-response behavioral chain*. At each learning step, if a robot has selected a direction for its next movement, then the components of W will be updated. Based on a series of updating, the weight of some motion direction will become more significant than the others, signifying that a motion in the respective direction has a higher likelihood of success than the rest, with respect to a specific stimulus. The conditioned (or empirically acquired) behavior patterns will serve as the basis for behavior selection in the robots of the same group.

6.2.1 *Formulation of Reinforcement Learning*

Suppose that a robot selects behavior B_i at time t when encountering stimulus $\int_k$. After the execution of the selected behavior, B_i, the robot learns based on its

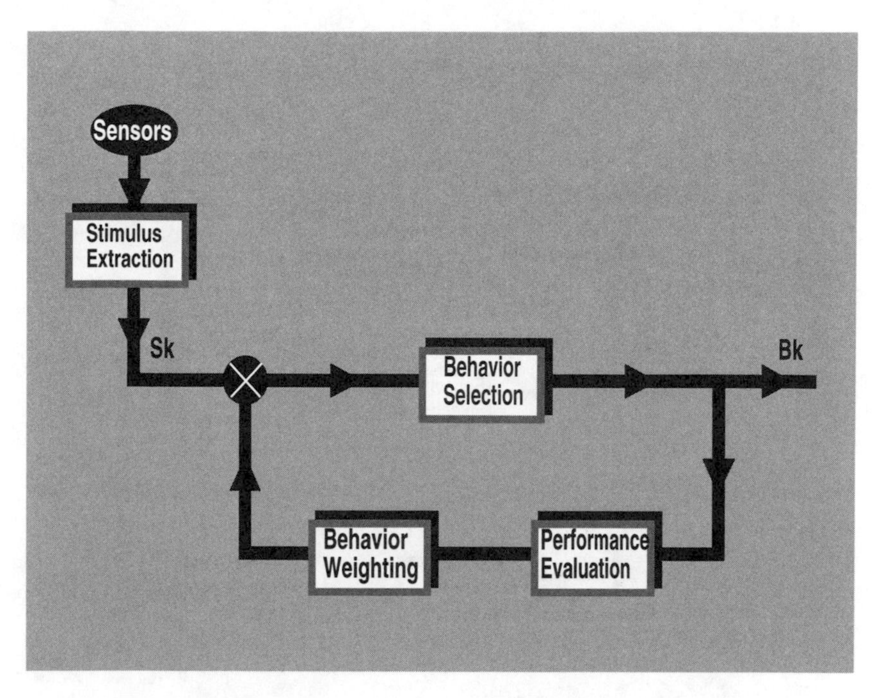

FIGURE 6.12. The behavioral conditioning mechanism in an autonomous robot.

local performance criteria. All group robots can share their learned knowledge since it is used as the common basis for behavior selection. Having learned from the executed behavior, the behavior weight vector will be updated as follows:

$$W_{f_k}^{t+1} = \texttt{normal}(\texttt{shape}(W_{f_k}^{t} + \Delta W)), \tag{6.12}$$

where ΔW is an increment vector. Operator `normal` normalizes the weight vector.

Operator `shape` is illustrated in Figure 6.14, where the updated weight vector passes through `function1`, and conditionally `function2`, before normalization. The definitions of `function1` and `function2` are given as follows:

- `function1`:

$$w_O = \begin{cases} 0, & \text{if } w_i < 0, \\ w_i, & \text{if } 0 \le w_i \le 1, \\ 1, & \text{if } w_i > 1. \end{cases} \tag{6.13}$$

- `function2`:

$$w_O = \frac{\alpha}{1 + e^{-w_i}} - \psi, \tag{6.14}$$

  where α and ψ are coefficients that affect the shape of the function. Figure 6.15 shows its shape when $\alpha = 3.9$ and $\psi = 1.9$.

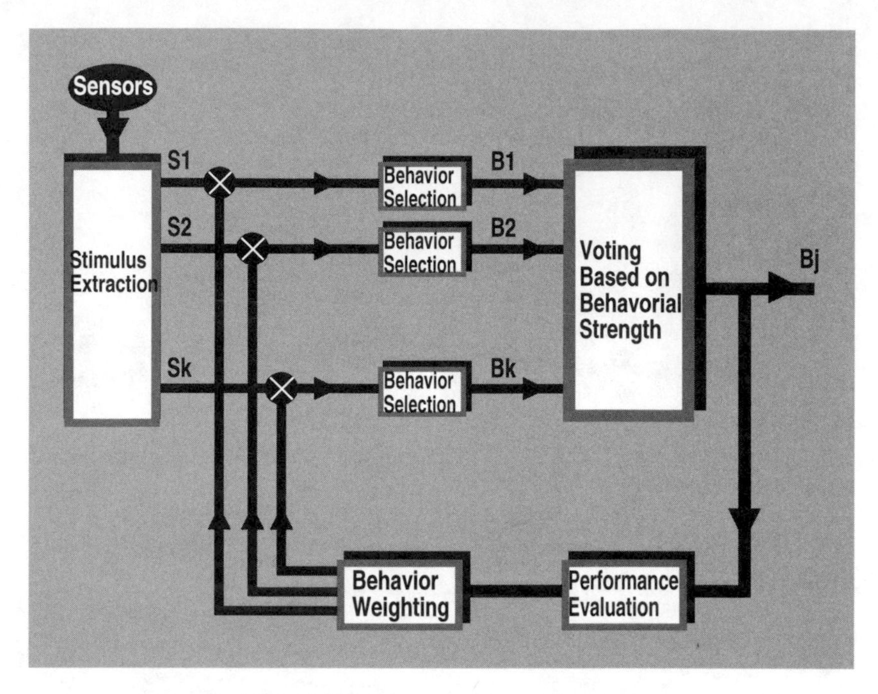

FIGURE 6.13. Voting-based behavior selection.

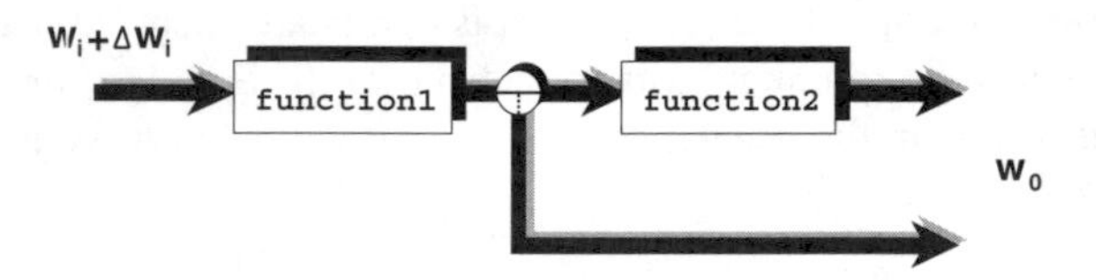

FIGURE 6.14. A schematic diagram of operator `shape` as used in the behavior learning mechanism.

Component j in weight increment vector ΔW is defined as follows:

$$\Delta w_j = \begin{cases} \delta \mid E(B_k), & \text{if } j = k, \\ 0, & \text{if } j \neq k, \end{cases} \tag{6.15}$$

where $E(B_k)$ is an evaluation of behavior B_k, and $\delta \in [-1, 1]$.

At time $t = 0$, component i of the behavior weight vector is computed as follows:

$$w_i^0 = \begin{cases} -1, & \text{if } B_i = -1, \\ \frac{1}{\mathcal{B}}, & \text{otherwise}, \end{cases} \tag{6.16}$$

where $\mathcal{B}$ denotes the number of feasible behaviors.

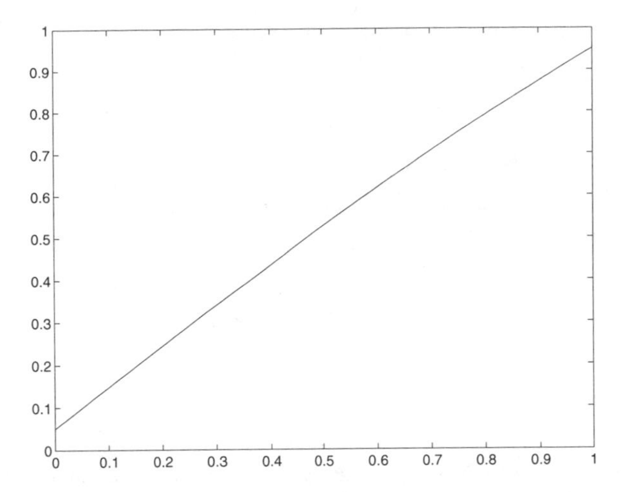

FIGURE 6.15. The shape of `function2` as in Eq. 6.14.

6.2.2 *Behavior Selection Mechanism*

Behavior selection refers to the mapping from a behavior weight vector to a behavior response. It can be expressed as follows:

$$\texttt{sel} : W \rightarrow B_k. \tag{6.17}$$

A group robot may choose one of the following two ways to select a behavior based on its behavior weight vector – a mapping from a current stimulus to a future behavior:

1. **Selection based on the probability distribution**: In this mechanism, a behavior response is selected based on the probability given in the behavior weight vector. The mechanism can be expressed as follows:

$$B_{\int_k} = B_k \mid_{\mathrm{P}(w_k)} . \tag{6.18}$$

2. **Selection based on the maximum weight**: In this mechanism, a behavior response will be chosen if it has the maximum probabilistic weight, that is:

$$B_{\int_k} = B_k \mid_{w_k = \texttt{max}(w_1, w_2, \cdots, w_{\mathcal{N}})} . \tag{6.19}$$

After the selection of a behavior, the robot moves in the selected direction with step d_0, as illustrated in Figure 6.11, where P^t and P^{t+1} denote the positions at time t and time $t + 1$, respectively. Here, we introduce operator `action` to represent the motion of robot i from P^t to P^{t+1} with selected behavior B_k and with step d_0, that is:

$$P_i^{t+1} = \texttt{action}(B_k,\ d_0,\ P_i^t). \tag{6.20}$$

It should be pointed out that there is an operation hidden in operator `action`. We name it `ret`. It maps from selected behavior B_k to movement sector β_j, that is:

$$\texttt{ret} : B_k \rightarrow \beta_j, \qquad k, j = 1, 2, \cdots, \mathcal{N}. \tag{6.21}$$

Based on the above definitions and notations, we can now define a mapping of behavior learning from an encountered stimulus to a reactive behavior, as follows:

$$\texttt{learn} : \mathcal{S} \rightarrow B_s. \tag{6.22}$$

6.3 Summary

Figure 6.16 presents an illustrative example to summarize behavior learning and selection as described in this chapter. In this example, we assume that each robot has eight short-range sensors and eight primitive behaviors.

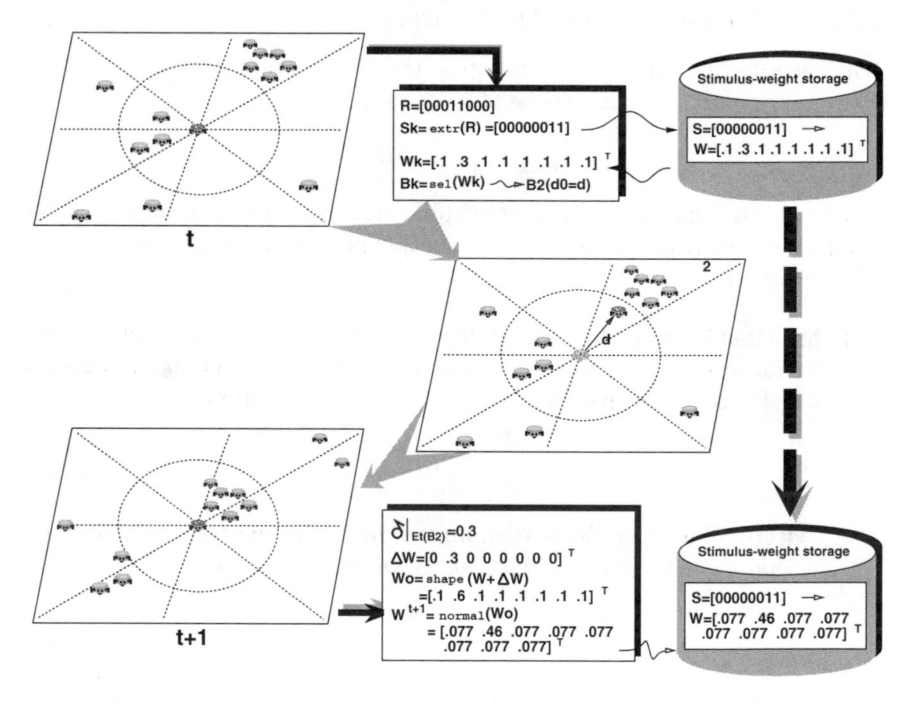

FIGURE 6.16. An illustrative example of behavior learning and selection.

7
Multi-Agent Reinforcement Learning: Results

The major result of this experiment was that some of my rats had babies.[1]

B. F. Skinner

Based on the design and formulation of a multi-agent reinforcement learning technique provided in the preceding chapter, this chapter will present the results of several experiments that examine the effectiveness of learning group behaviors in autonomous robots and highlight some of the key observations obtained from these experiments as well as their underlying conditions. The aim is to demonstrate the interrelationships between the local autonomy of individual robots and their emergent spatial properties as a result of the dynamic interactions between robots of the same class and their environment that may involve robots of other classes.

[1]A case history in scientific method. *American Psychologist*, 11, 1956, p 221-233.

We will elaborate on our experimental procedures for investigating group behaviors, covering the experimental conditions involved and the experimental measurements employed. This is followed by a summary of the experimental results with respect to the proposed measurement variables.

7.1 Measurements

In order to quantitatively characterize the spatial conditions of an environment and the behavior performance of group robots, we will introduce two measurements: stimulus frequency and behavior selection frequency.

7.1.1 *Stimulus Frequency*

Stimulus frequency refers to the frequency at which group robots encounter a specific stimulus. Let us define a set $\mathcal{J}(\int_k, t)$ to contain all the group robots that encounter stimulus $\int_k$ at time t, that is:

$$\mathcal{J}(\int_k, t) = \{P_i \mid \int_i^t = \int_k, \qquad i = 1, 2, \cdots, M\}, \tag{7.1}$$

where M denotes the number of robots in a group. Other symbols refer to the same meanings as those defined in the preceding chapter.

Next, we define the number of times that the group robots encounter stimulus $\int_k$ at time t as follows:

$$H_{\int_k}^t = \texttt{size}(\mathcal{J}(\int_k, t)), \tag{7.2}$$

where operator `size` returns the number of elements in $\mathcal{J}(\int_k, t)$.

Figure 7.1 presents an example of measurement $H_{\int_k}^t$. The upper subfigure shows the changes of $H_{\int_k}^t$ over time, with respect to a group of robots called `RANGER` that encounters `STIMULUS` $=$ 00000000. The lower subfigure provides the corresponding weight changes associated with the primitive behaviors of the `RANGER` robots.

The naming convention for the example given in Figure 7.1 as well as other examples in the following sections is summarized in Table 7.1[2].

7.1.2 *Behavior Selection Frequency*

Behavior selection frequency is another important measurement that we will use to quantitatively describe a multi-agent robotic system. More specifically, we let

---

[2]The remaining letters in a case name, **Cx**Ry**Wz, denote the x th definition for the weight increment (ΔW) in behavior weight updating, the y number of `RANGER` robots, and the z number of `WILD` robots, respectively.

	1	2	3	4
B: Target motion characteristic	learning to move away	stationary	random motion	
D: Target distribution, wall-closure	decentralized, no wall	decentralized, wall	centralized, no wall	centralized, wall
L: Learning, selection mechanism	smoothed weights, by probability	smoothed weights, by maximum weight	original weights, by probability	original weights, by maximum weight
S: Sensor range	short	long		

TABLE 7.1. The naming convention as used in multi-agent reinforcement learning case studies (©1998 IEEE).

$\Gamma(B_k, t)$ denote a set of group robots that execute behavior B_k at time t, which is formally expressed as follows:

$$\Gamma(B_k, t) = \{P_i \mid B_i^t = B_k, \qquad i = 1, 2, \cdots, M\}, \tag{7.3}$$

where M denotes the number of robots in the group. Thus, the behavior selection frequency of B_k, that is, the number of times that behavior B_k is selected by the group robots, as denoted by $\mathtt{ID}_{B_k}^t$, can be readily derived by performing the following operation:

$$\mathtt{ID}_{B_k}^t = \mathtt{size}(\Gamma(B_k, t)). \tag{7.4}$$

Figure 7.2 gives an example plot of behavior selection frequency $\mathtt{ID}_{B_k}^t$ for 8 primitive behaviors. From the figure, it can be noted that the most frequently selected behaviors are behaviors 1 and 2.

7.2 Group Behaviors

In this section, we will discuss how group behaviors can be produced effectively based on the earlier described multi-agent reinforcement learning. The scenario that we will focus on consists of two distinct groups of robots, `RANGER` robots and `WILD` robots, in a bounded rectangular environment. The `WILD` robots can only perform simple behaviors, such as `randommotion` and `escaping`. The `RANGER` robots, on the other hand, are supposed to find out how to collectively

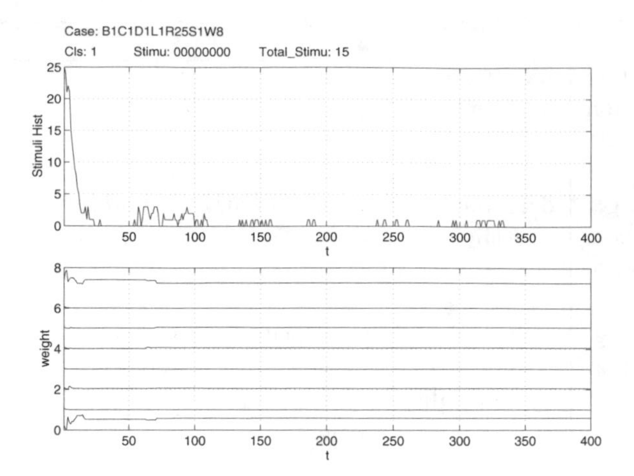

FIGURE 7.1. An example of measurement $H^t_{f_k}$.

search, approach, follow, and surround the WILD robots by means of multi-agent reinforcement learning.

7.2.1 Collective Surrounding

In our experiments, a group of 25 simulated autonomous robots, as denoted by the $*$ symbols in Figure 7.3(a), will be used. The individual robot in this group can receive time-varying stimuli from its environment, e.g., the change in the number of WILD robots within its neighboring region before and after the execution of a selected behavior. The WILD robots are marked with the $\bullet$ symbols in the figure. Without loss of generality, the task of the RANGER robots is to surround the set of 8 WILD robots placed within a 10×10 area.

Initially, the RANGER robots are randomly distributed in their environment without *a priori* knowledge about their environment and desirable motion behaviors, whereas the WILD robots are placed together as a crowd, as shown in Figure 7.3(a). In this case, the RANGER robots are found to be capable of quickly developing a group behavior that enables them to move toward the WILD crowd. From Figures 7.3(b) and (c), we note that all RANGER robots move directly toward the WILD robots.

From Figure 7.4(a), we realize that all RANGER robots initially encounter STIMULUS1. With respect to this stimulus, the robots select behavior $B1$ as their response. Although we can see in Figure 7.3(b) that there is one robot moving against the WILD crowd at the beginning, after the evaluation to the selected behavior, its corresponding weight decreases. At the same time, other behavior weights increase. That is why we find in Figure 7.4(a) that the weight of behavior $B1$ reaches 1 quickly. Due to the knowledge sharing mechanism, all robots will select behavior $B1$ when encountering STIMULUS1. This can be denoted as follows:

$$\texttt{learn(STIMULUS1)} \rightarrow \texttt{B}_s = \texttt{B1}. \qquad (7.5)$$

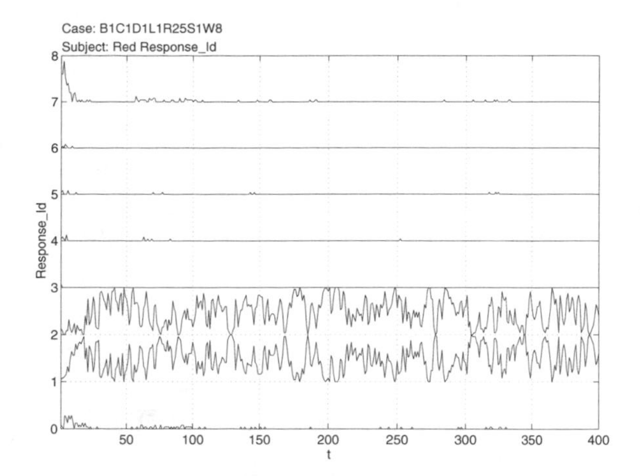

FIGURE 7.2. An example plot of $\mathtt{ID}^t_{B_k}$.

This learning result can also be expressed as the following rule:

$$\begin{array}{l} \textit{Rule RA1}: \\ \qquad \forall i \in \mathtt{RANGER},\ \text{if } \int_i = \mathtt{STIMULUS1} \\ \qquad\qquad \Downarrow \\ \qquad P_i^{t+1} = \mathtt{action}(B1,\ d_0,\ P_i^t), \end{array} \tag{7.6}$$

where symbols $\int_i$, P_i, and d_0 refer to the same meanings as those defined in the preceding chapter. An illustration of this rule is given in Figure 7.5(a).

After performing operation `ret`, all robots will move in a direction where most `WILD` robots can be found. This trend is quite evident from Figure 7.3(b).

Referring to Figure 7.3(b) and Figure 7.4(b), we can note that as more and more `RANGER` robots surround the `WILD` group, some of the `RANGER` robots will encounter `STIMULUS2`. Similarly, as a result of learning, behavior $B8$ (corresponding to region j after mapping: $\beta_j = \mathtt{ret}(B8)$) will be selected as their moving direction, as indicated in Figure 7.4. Although other moving directions may have been selected before reaching the stable state (after about 20 steps), the behavior corresponding to this direction is more positively reinforced and hence more likely to be selected than others during the group learning based on Eq. 6.15. As a result, the response to `STIMULUS2` in this case can be acquired and expressed as follows:

$$\begin{array}{l} \textit{Rule RB1}: \\ \qquad \forall i \in \mathtt{RANGER},\ \text{if } \int_i = \mathtt{STIMULUS2} \\ \qquad\qquad \Downarrow \\ \qquad P_i^{t+1} = \mathtt{action}(B8,\ d_0,\ P_i^t). \end{array} \tag{7.7}$$

Figure 7.5(b) illustrates this rule. Figure 7.3(d) shows the resulting stable state of collective surrounding.

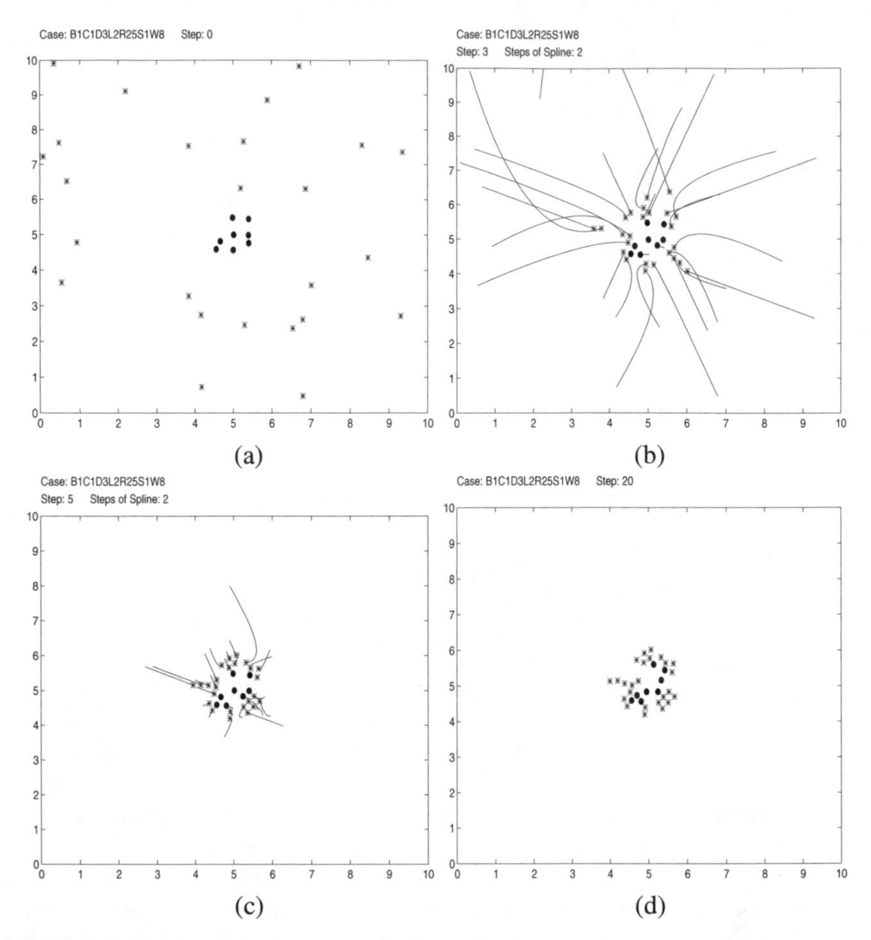

FIGURE 7.3. (a) The initial distribution of robots in an environment. (b) (c) All `RANGER` robots move directly toward the `WILD` robots. (d) A stable distribution is reached at step 20.

7.2.2 *Cooperation among* RANGER *Robots*

The next question that we are interested in is whether the `RANGER` robots can acquire a cooperative behavior in order to effectively surround the `WILD` group. From our experiments, it can be observed that as a result of learning, the `RANGER` robots will move to positions around the `WILD` robots. For instance, when a `RANGER` robot finds that a `WILD` robot is surrounded by fewer `RANGER` robots, it will join the shorthanded group. Also, some `RANGER` robots will cooperatively push one or more `WILD` robots.

7.2.2.1 Moving away from Spatially Cluttered Locations

From Figure 7.6, we note that a `RANGER` robot moves away from the back of other `RANGER` robots and then finds itself a new position. This behavior is induced from the learning process as shown in Figure 7.7. We can find that the last robot that receives STIMULUS1 finds its new optimal position at step 184. Although this robot previously reached a locally optimal state about 50 steps ago, it was at step 177 when it selected a randomizedmotion behavior that caused the robot to move away from this locally optimal state. Thus, before the system reaches its globally optimal state, the selection of some *temporally worse* (or mutated) behaviors by the individual robots can be helpful for the system to reach a globally optimal state. The observation is summarized in rule *RC1* below (see Figure 7.8):

$$\begin{aligned} &\textit{Rule RC1}: \\ &\qquad \forall i \in \texttt{RANGER}, \text{if } f_i = \texttt{STIMULUS1}, \ \exists T, \ \exists \Delta T, \ \forall t \in [t_0, \ t_0 + T], \\ &\qquad \tfrac{dw_i}{dt} = 0, \text{ and } T > \Delta T, \\ &\qquad \Downarrow \\ &\qquad P_i^{t+1} = \texttt{action}(B_m, \ d_0, \ P_i^t), \\ &\qquad \text{where } B_m = \texttt{mutation}(B1). \end{aligned} \tag{7.8}$$

Operator mutation returns a randomly mutated behavior based on a primitive behavior.

The multi-agent system reaches its stable state as all the RANGER robots are governed by rule *RB1* as shown in Eq. 7.7, which is also reflected from Figure 7.7(b).

7.2.2.2 Changing a Target

Figure 7.9(a) shows a quasi-stable state where some RANGER robots have not reached their optimal positions even though all of them are surrounding the WILD robots. By applying *RC1* as expressed in Eq. 7.8, the system will be able to transition to its globally optimal state. Figure 7.9(b) presents a situation where a RANGER jumps away from one crowd and moves to a less surrounded WILD robot.

Also from Figure 7.10, we can find that some RANGER robots with STIMULUS1 are gradually leaving the RANGER crowd to approach other WILD robots. After step 50, all RANGER robots are with the WILD robots. Thereafter, a stable state is maintained by rule *RB1* as expressed in Eq. 7.7.

7.2.2.3 Cooperatively Pushing Scattered Objects

Figures 7.11 and 7.12 present two cases where a WILD robot is gradually being pushed to the left. The three coordinated RANGER robots push the WILD robot at the same pace in the same direction until it is obstructed by another RANGER robot. Figure 7.13 plots the process of acquiring a response to STIMULUS2. The resulting behavior is referred to as rule *RB1*.

7.2.2.4 Collective Manipulation of Scattered Objects

In the above cases on the cooperation among the RANGER robots, we note that the robots can jump away from a crowd and find better positions, but they will never stay at common locations. In other words, they tend to be spatially distributed around the WILD robots. This phenomenon is created by the robots based on the coupling of the above three rules, that is:

$$\begin{aligned}&\textit{Rule}: \\ &\qquad \texttt{sharing} = RA1 \otimes RB1 \otimes RC1.\end{aligned} \tag{7.9}$$

where $\otimes$ means *coupling*.

Figure 7.14 presents a typical example of spatial distribution produced by the coupling of the acquired *surrounding* (rule *RA1*), *pursuing* (rule *RB1*), and *mutation* (rule *RC1*).

7.2.3 Concurrent Learning in Different Groups of Robots

7.2.3.1 Concurrent Learning in *Predator* and *Prey*

Figure 7.15 shows a spatial distribution of the WILD robots after 4 steps. Figures 7.16 and 7.17 provide two snapshots of collective pushing by the RANGER robots at steps 9 and 19, respectively. Figures 7.18 to 7.24 present the processes of learning behavior responses with respect to various stimuli in two different robot groups. In this case, the RANGER robots often encounter the stimuli of STIMULUS1, STIMULUS2, and STIMULUS3, as observed from Figures 7.18 to 7.20. The result of learning as shown in Figure 7.18 is consistent with rule *RA1* as defined in Eq. 7.6. On the other hand, the result shown in Figure 7.19 agrees well with rule *RB1* as defined in Eq. 7.7, corresponding to the group behavior of *following* or *pushing*. After the mapping of ret, the relative direction (the eighth sector to STIMULUS1 and the coupling of the fifth and the seventh sectors to STIMULUS2) determines their respective motion sector. In this case, the acquired responses to STIMULUS3, STIMULUS4, STIMULUS7, and STIMULUS8 are expressed as follows (see Figure 7.25):

$$\begin{aligned}&\textit{Rule RB2}: \\ &\qquad \forall i \in \texttt{RANGER},\ \text{if } \int_i = \texttt{STIMULUS3} \\ &\qquad\qquad \Downarrow \\ &\qquad P_i^{t+1} = \texttt{action}(B1 \otimes B8,\ d_0,\ P_i^t).\end{aligned} \tag{7.10}$$

$$\begin{aligned}&\textit{Rule RB3}: \\ &\qquad \forall i \in \texttt{RANGER},\ \text{if } \int_i = \texttt{STIMULUS4} \\ &\qquad\qquad \Downarrow \\ &\qquad P_i^{t+1} = \texttt{action}(B7,\ d_0,\ P_i^t).\end{aligned} \tag{7.11}$$

$$\begin{aligned}&\textit{Rule RB4}: \\ &\qquad \forall i \in \texttt{RANGER},\ \text{if } \int_i = \texttt{STIMULUS7} \\ &\qquad\qquad \Downarrow \\ &\qquad P_i^{t+1} = \texttt{action}(B1 \otimes B5,\ d_0,\ P_i^t).\end{aligned} \tag{7.12}$$

Rule RB5 :

$$\forall i \in \texttt{RANGER},\ \text{if}\ \int_i = \texttt{STIMULUS8}$$
$$\Downarrow \tag{7.13}$$
$$P_i^{t+1} = \texttt{action}(B6 \otimes B7,\ d_0,\ P_i^t).$$

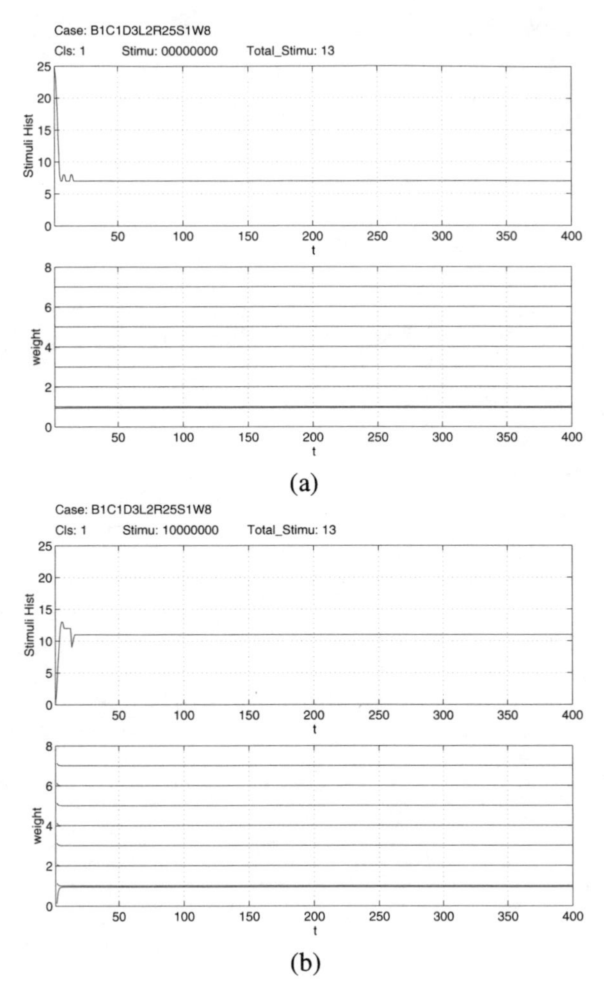

FIGURE 7.4. The histograms of (a) STIMULUS1 and (b) STIMULUS2, along with their corresponding behavior weights for the RANGER group.

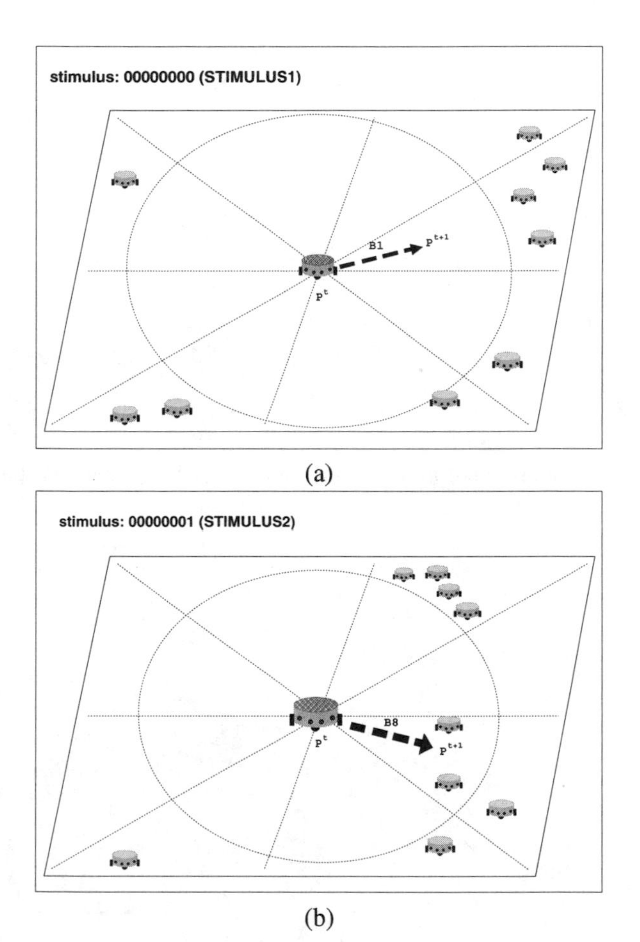

FIGURE 7.5. An illustration of rules (a) *RA1* and (b) *RB1*.

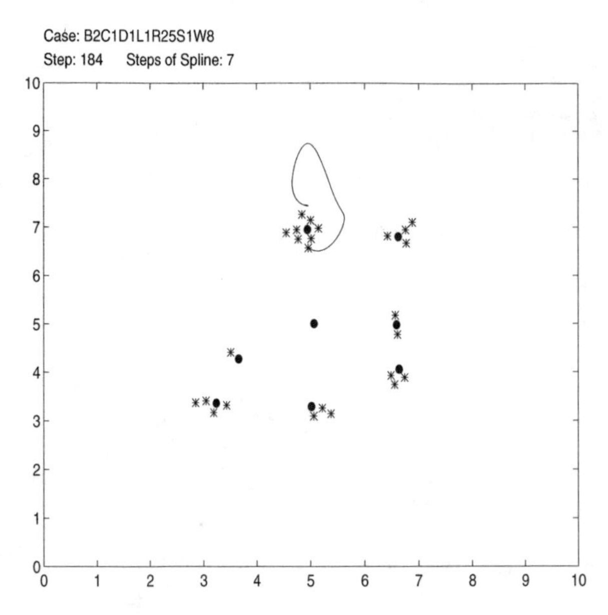

FIGURE 7.6. A RANGER robot moves away from the back of other RANGER robots to a better position.

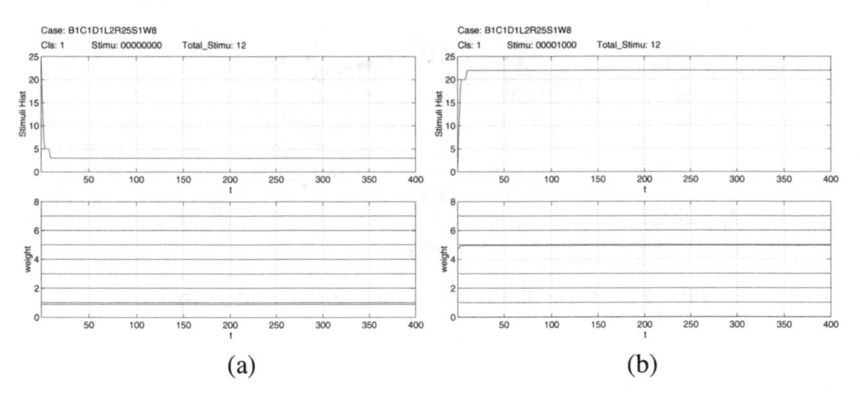

FIGURE 7.7. The histograms of (a) STIMULUS1 and (b) STIMULUS2, along with their corresponding behavior weights the RANGER group in the case where the WILD group is stationary.

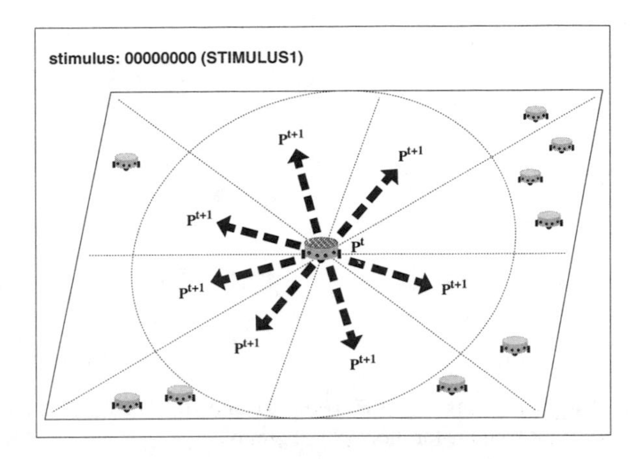

FIGURE 7.8. An illustration of rule *RC1*.

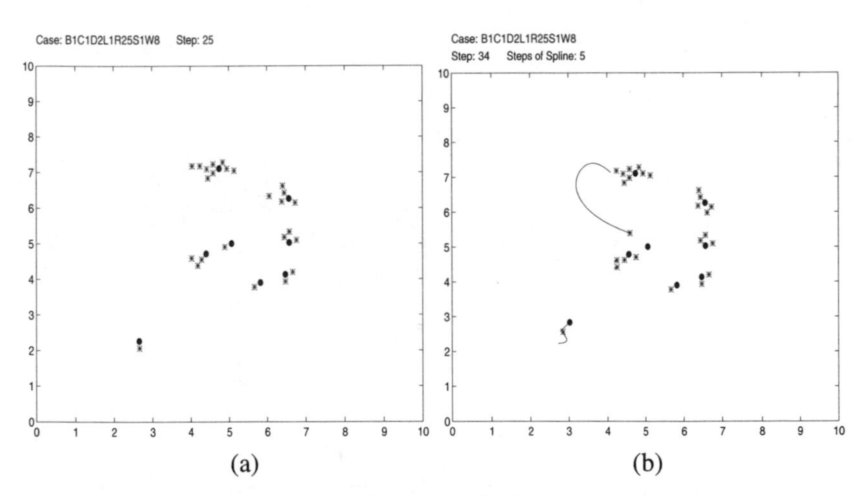

FIGURE 7.9. (a) A quasi-stable distribution at step 25. (b) A RANGER robot jumps away from a RANGER crowd to find other less surrounded WILD robots.

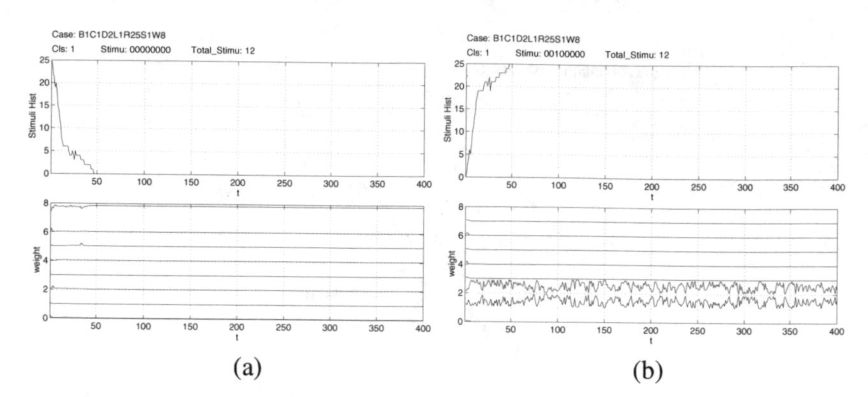

FIGURE 7.10. The histograms of (a) STIMULUS1 and (b) STIMULUS2, along with the corresponding behavior weights for the RANGER group.

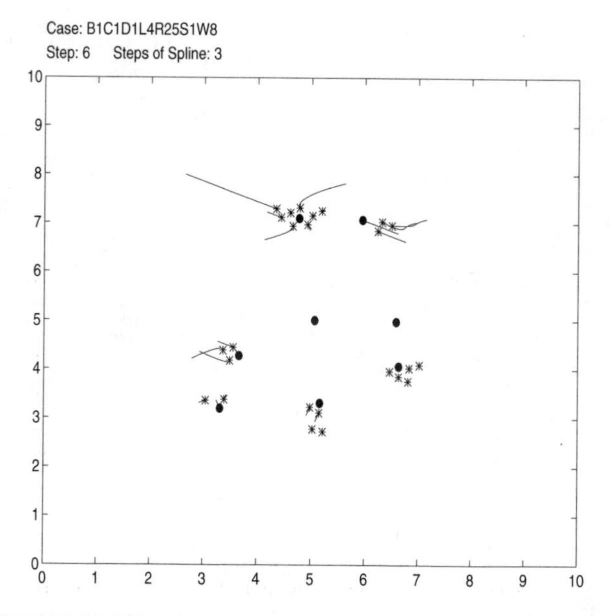

FIGURE 7.11. Three RANGER robots cooperatively push a WILD robot.

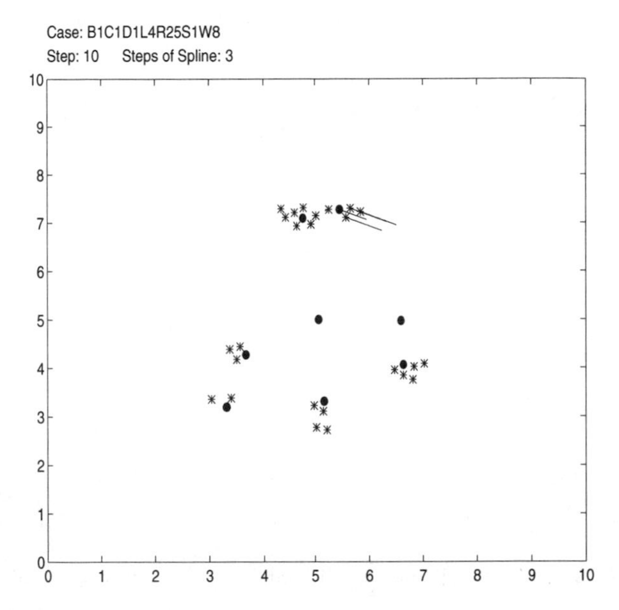

FIGURE 7.12. Three RANGER robots cooperatively push a WILD robot.

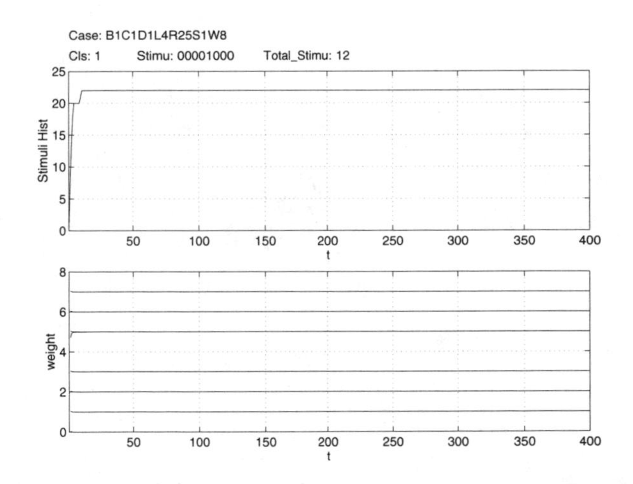

FIGURE 7.13. The process of learning a response to STIMULUS2.

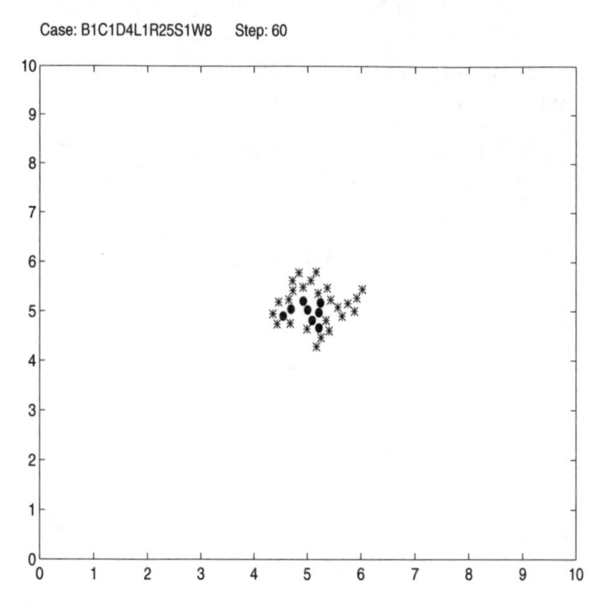

FIGURE 7.14. A spatial distribution of the RANGER robots observed during collective surrounding.

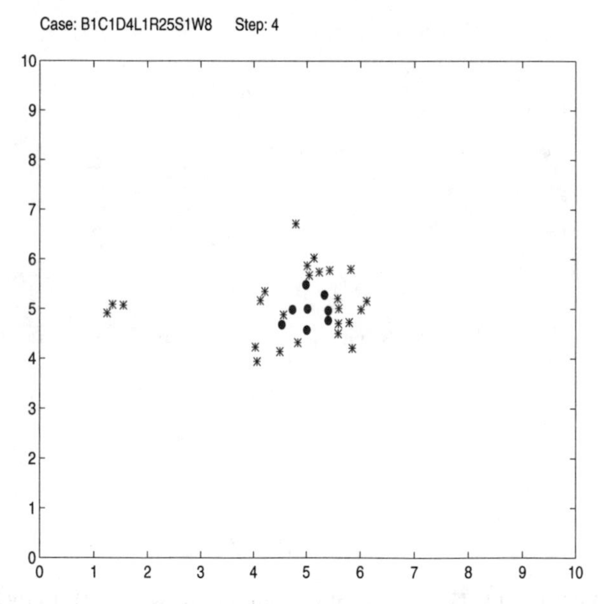

FIGURE 7.15. The loose distribution of WILD robots after 4 steps in the case of B1C1D4L1R25S1W8.

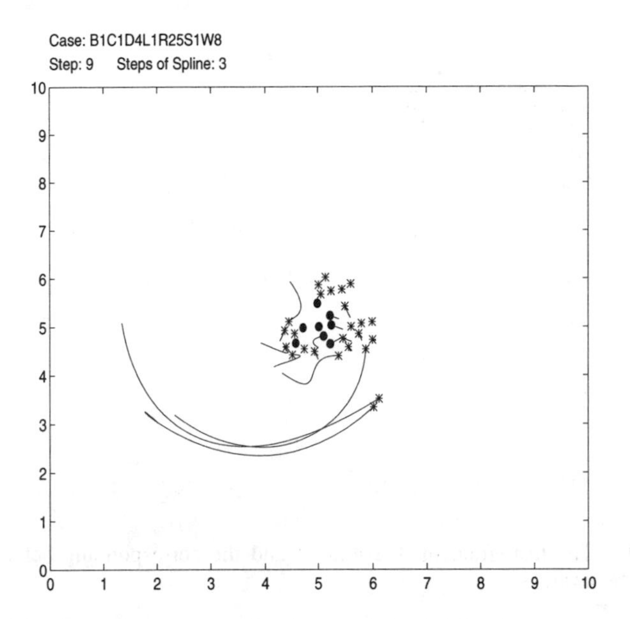

FIGURE 7.16. The RANGER robots push the WILD robots to move.

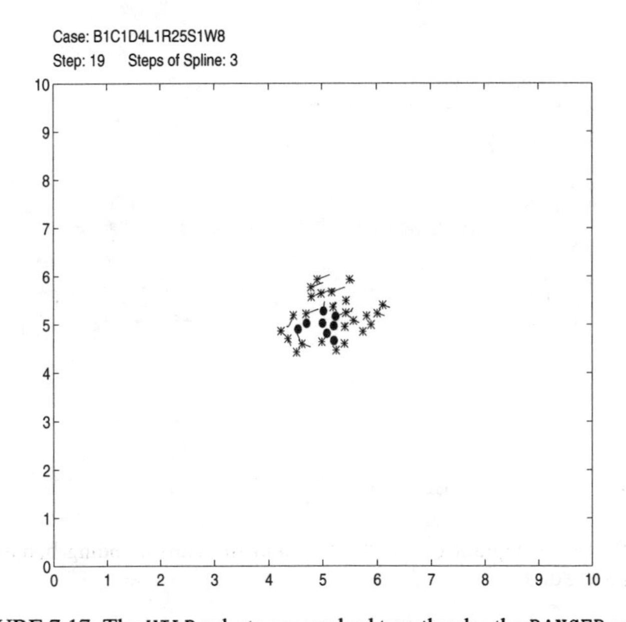

FIGURE 7.17. The WILD robots are pushed together by the RANGER robots.

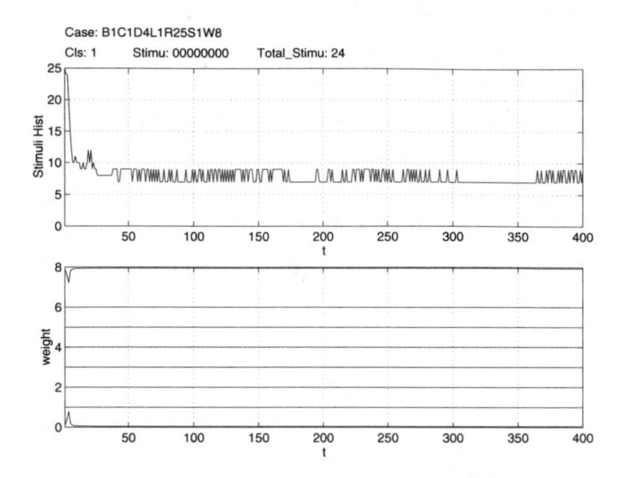

FIGURE 7.18. The histogram of STIMULUS1 and the corresponding behavior weights acquired by the RANGER.

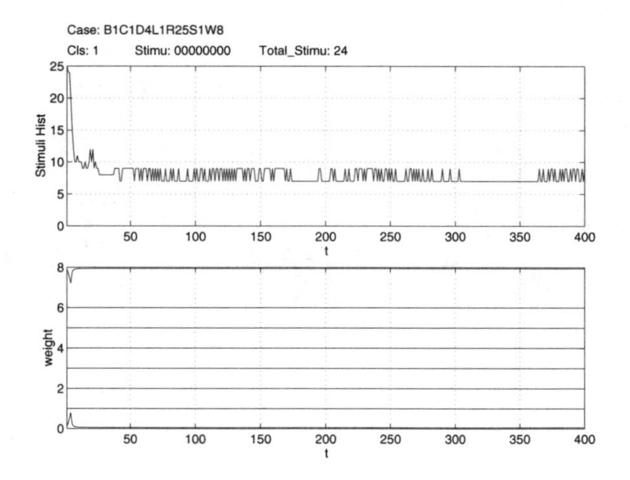

FIGURE 7.19. The histogram of STIMULUS2 and the corresponding behavior weights acquired by the RANGER.

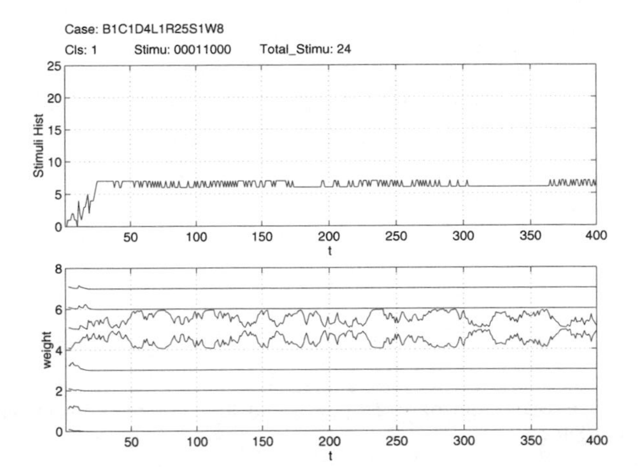

FIGURE 7.20. The histogram of STIMULUS3 and the corresponding behavior weights acquired by the RANGER.

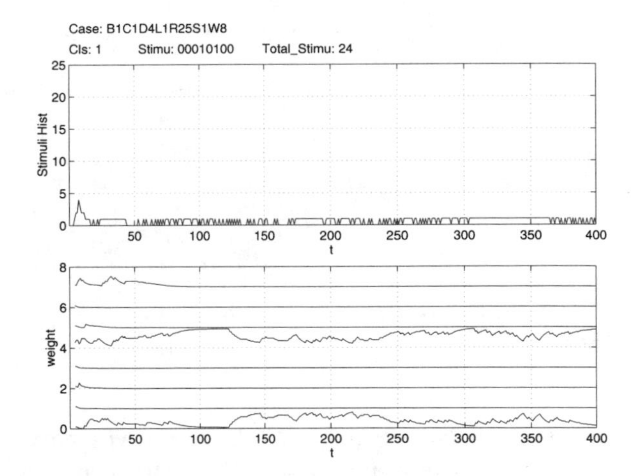

FIGURE 7.21. The histogram of STIMULUS4 and the corresponding behavior weights acquired by the RANGER.

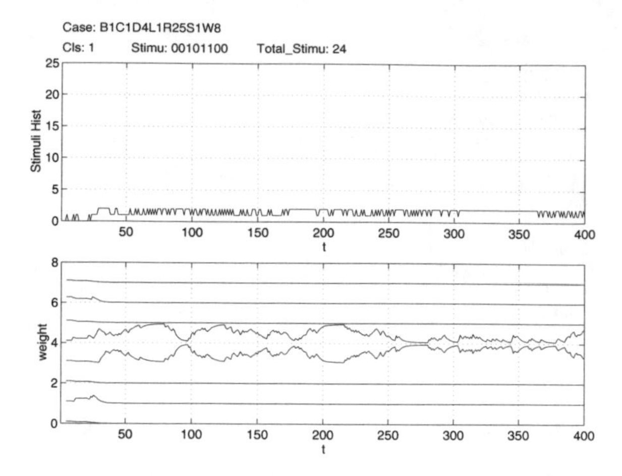

FIGURE 7.22. The histogram of STIMULUS8 and the corresponding behavior weights acquired by the RANGER.

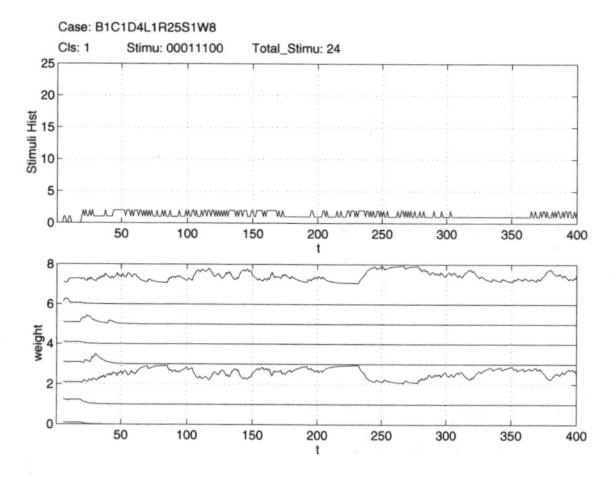

FIGURE 7.23. The histogram of STIMULUS7 and the corresponding behavior weights acquired by the RANGER.

These learning results indicate that the RANGER robots will move closer to the regions in or near which there is/are some WILD robot(s) and hence produce a propulsive force to the WILD.

Similarly, it can be realized from Figure 7.24 that as a WILD robot encounters STIMULUS8, it is very likely to select the first and the fifth sector. This means the WILD robot will try to escape when the RANGER pursues, as stated in rule *RD1* below (see Figure 7.26):

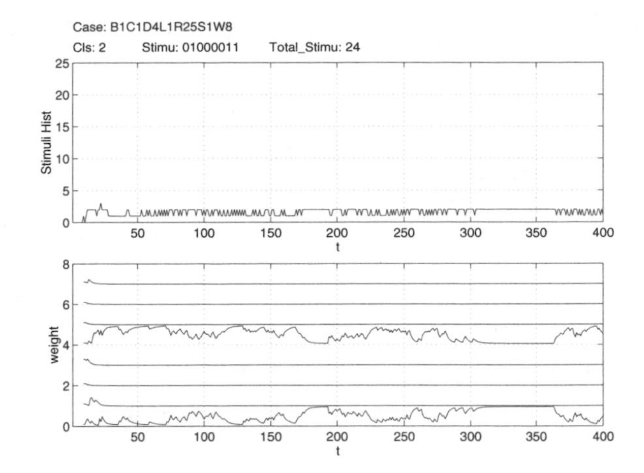

FIGURE 7.24. The histogram of STIMULUS8 and the corresponding behavior weights acquired by the WILD.

$$\begin{aligned} &\textit{Rule RD1}: \\ &\qquad \forall i \in \texttt{WILD},\ \text{if } \int_i = \texttt{STIMULUS8} \\ &\qquad\qquad \Downarrow \\ &\qquad P_i^{t+1} = \texttt{action}(B2,\ d_0,\ P_i^t). \end{aligned} \tag{7.14}$$

It should be pointed out that the number of times that the WILD encounters STIMULUS8 may be just once or twice, since many WILD robots are surrounded in the center by other WILD robots.

From the macroscopic point of view, what will be observed as a result is that a crowd of the WILD robots that are surrounded by the RANGER will move closer and closer to each other, as shown in Figure 7.17.

7.2.3.2 Chasing

Figure 7.27 presents a case where some RANGER robots have learned to chase the WILD robots. Figure 7.28 shows the histogram and the corresponding behavior weights acquired by the two robot groups. As shown in Figure 7.28(a), after a RANGER robot encounters STIMULUS2 (that is, some WILD robots are found in the third sector), it will be very likely to select the second or third sector as its motion direction, as governed by rule *RB1* of Eq. 7.7. As a result, it stays together with the caught WILD robot(s).

Figure 7.27 shows an example where several WILD robots are being chased by the RANGER robots until they encounter STIMULUS2. From Figure 7.28(b), we can find that in this case, the WILD is very likely to move to the third or the fourth sector. Because the WILD finds that there are RANGER robots in the seventh sector, it will select the behavior of moving against the RANGER; that is to say, it will try to escape from the RANGER. The acquired rule of escaping for a WILD robot in this case can be expressed as follows (see Figure 7.29):

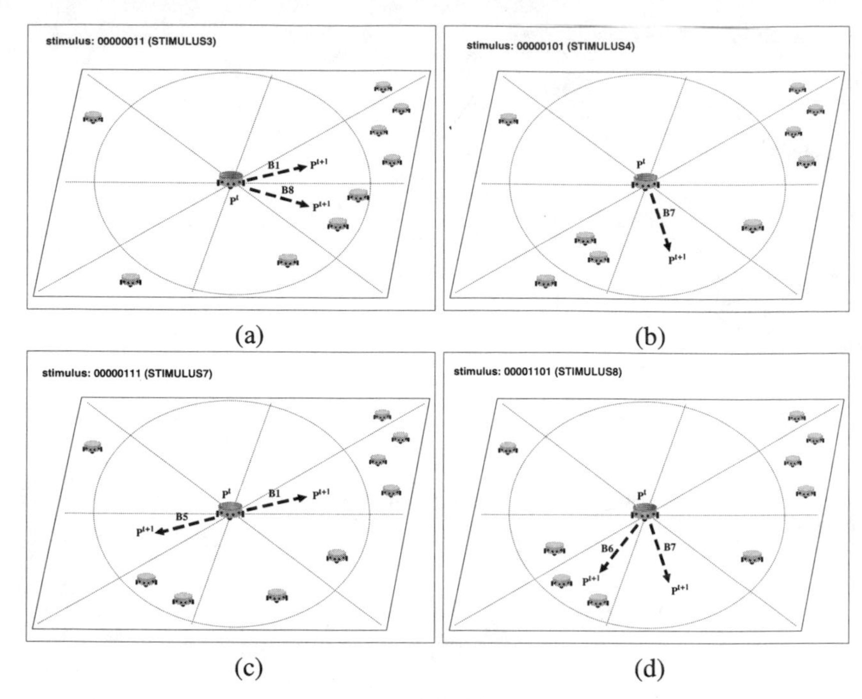

FIGURE 7.25. An illustration of rules (a) *RB2*, (b) *RB3*, (c) *RB4*, and (d) *RB5*.

$$\textit{Rule RD2}: \quad \begin{array}{c} \forall i \in \texttt{WILD}, \text{ if } \int_i = \texttt{STIMULUS2} \\ \Downarrow \\ P_i^{t+1} = \texttt{action}(B4,\ d_0,\ P_i^t). \end{array} \tag{7.15}$$

7.2.3.3 Escaping from a Surrounding Crowd

As can be noted from Figure 7.30(a), a WILD robot will try to escape from being surrounded by the RANGER robots. This is evident from the corresponding behavior weights given in Figure 7.30(b). Based on group learning, the WILD robots will select the fifth or the seventh sector as its motion direction when they identify the distribution around them as 01110101. In other words, the WILD robots will escape from being surrounded by the RANGER. The rule of escaping for a WILD robot in such a case can be expressed as follows (see Figure 7.31):

$$\textit{Rule RD3}: \quad \begin{array}{c} \forall i \in \texttt{WILD}, \text{ if } \int_i = \texttt{STIMULUS25} \\ \Downarrow \\ P_i^{t+1} = \texttt{action}(B4 \otimes B6,\ d_0,\ P_i^t). \end{array} \tag{7.16}$$

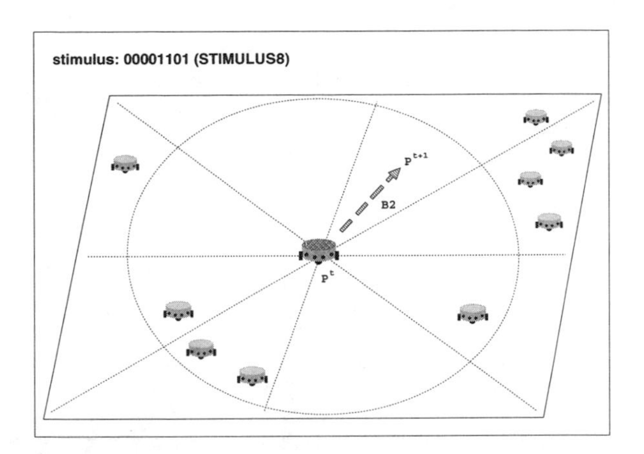

FIGURE 7.26. An illustration of rule *RD1*.

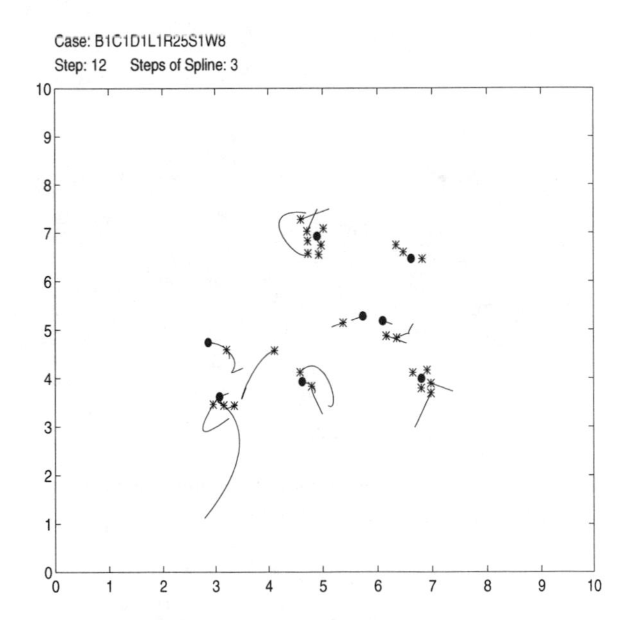

FIGURE 7.27. The RANGER robots follow the WILD robots.

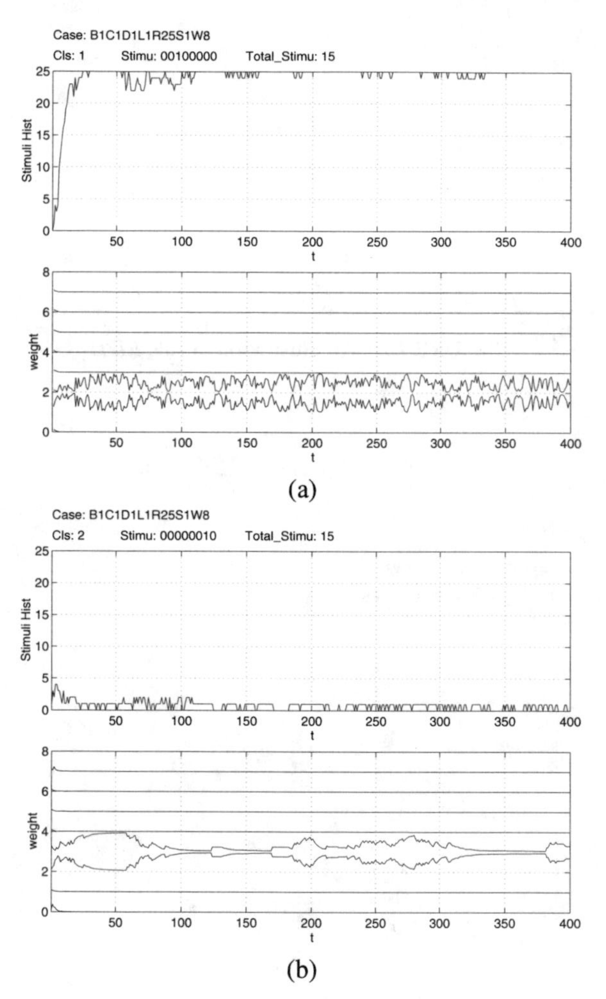

FIGURE 7.28. The histogram and corresponding behavior weights for (a) `STIMULUS2` in the `RANGER` group and (b) `STIMULUS2` in the `WILD` group.

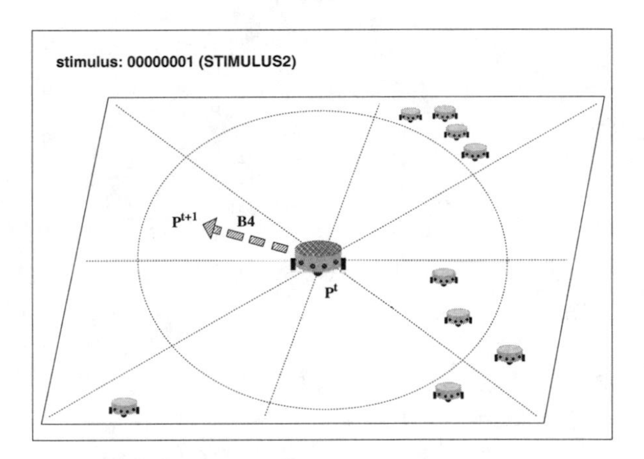

FIGURE 7.29. An illustration of rule *RD2*.

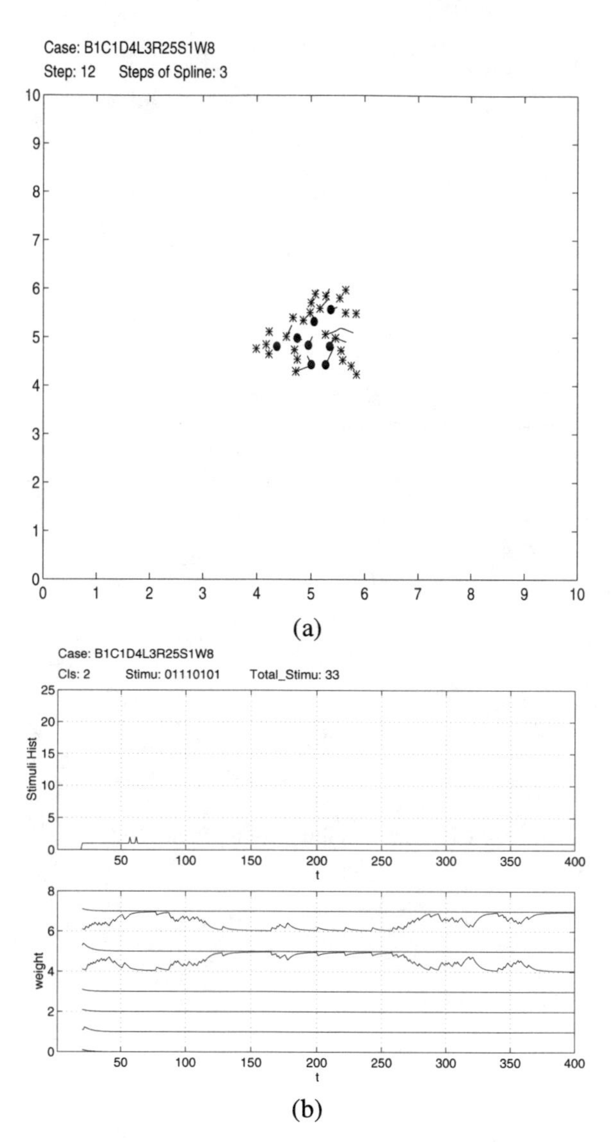

FIGURE 7.30. (a) A WILD robot tries to escape from being surrounded by the RANGER robots. (b) The histogram of STIMULUS25.

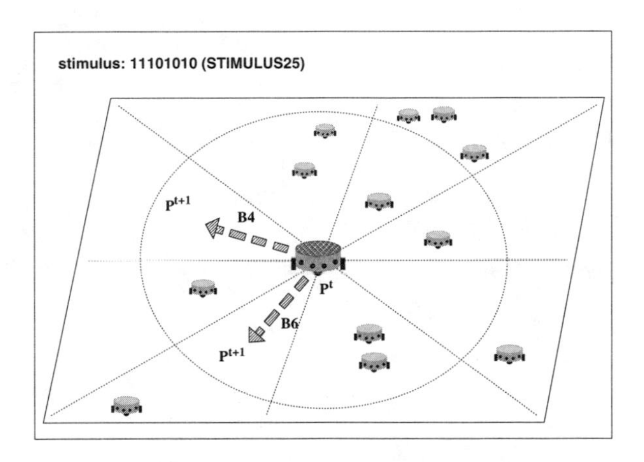

FIGURE 7.31. An illustration of rule *RD3*.

8

Multi-Agent Reinforcement Learning: What Matters?

A bridge, under its usual conditions of service, behaves simply as a relatively smooth level surface on which vehicles can move. Only when it has been overloaded do we learn the physical properties of the materials from which it is built.[1]

Herbert A. Simon

The preceding chapters described a multi-agent reinforcement learning technique and various reactive behaviors acquired by group robots using such a technique. This chapter will further discuss the important factors that can influence multi-agent learning. These factors include group sensors, initial distribution, behavior selection mechanism, and motion mechanism used by a `WILD` group.

[1] *The Sciences of the Artificial*, The MIT Press, Cambridge, Massachusetts, 1996 (3rd ed.), p 13.

8.1 Collective Sensing

Long-range sensors can cover an entire environment. However, with this type of sensors, it is sometimes difficult to reflect some slight changes in the positions of other robots in the sensory data. As a result, such sensors may be insensitive to the local motions of the robots. Short-range sensors, on the other hand, divide the surrounding area into two regions separated by a circle with the radius of sensing length. The spatial arrangements in each of the two regions can be dealt with, respectively. In other words, with the second type of sensors, a finer, more accurate description of the environment can be achieved. In order to demonstrate this point, let us take a look at some examples:

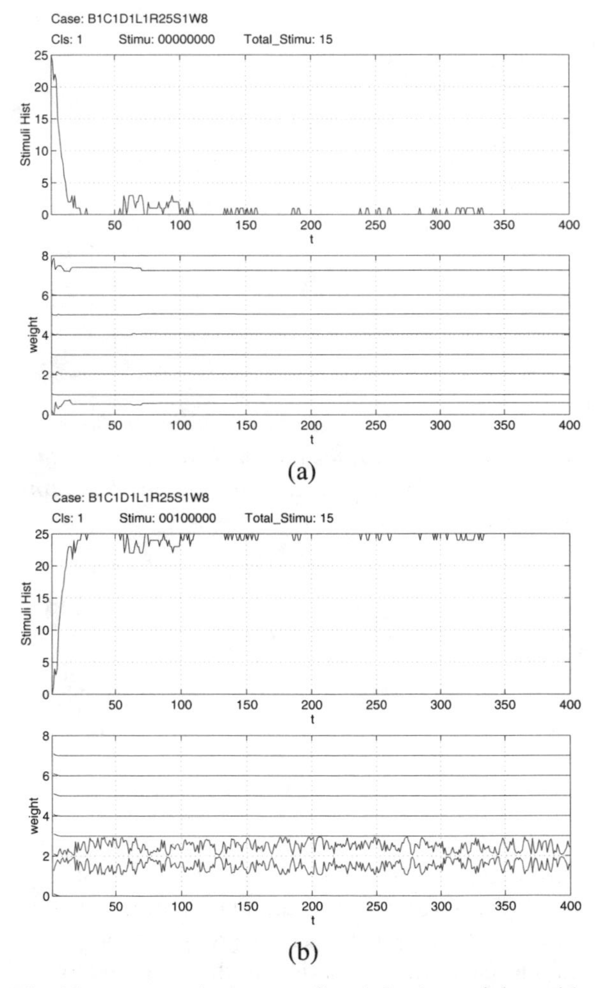

FIGURE 8.1. The histogram and corresponding behavior weights with respect to (a) STIMULUS1 (b) STIMULUS2 in the RANGER group in the case of B1C1D1L1R25S1W8.

1. **Case B1C1D1L1R25S1W8 (short-range) vs. Case B1C1D1L1R25S2W8 (long-range)**: All the conditions in these two cases are the same except that two different types of sensors are used. Figure 8.1 shows the histograms of STIMULUS1 and STIMULUS2 as extracted by the RANGER group and their corresponding behavior weights. From the figure, we can observe that in the case of B1C1D1L1R25S1W8, after several steps of group learning, two directions are identified for reactive motion with respect to each stimulus. After 30 steps, the RANGER robots start to encounter STIMULUS2 and the system reaches a stable state. This is governed by rules *RA1* and *RA2*. The macroscopic spatial distances between the RANGER robots, between the RANGER and the WILD robots, and the selected sectors at each step are given in Figures 8.2 (a), (b), and (c), respectively. From these figures, we can note that the system is stable in a macroscopic sense.

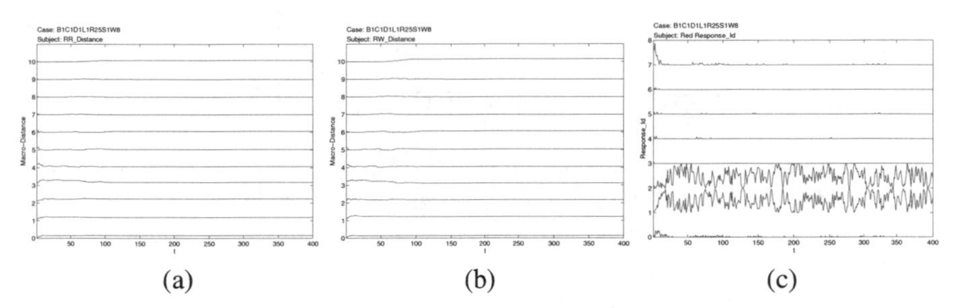

(a) (b) (c)

FIGURE 8.2. The distributions of the spatial distances (a) between two RANGER robots and (b) between a RANGER and a WILD robot. (c) Behaviors selected by the RANGER robots.

As a comparison, Figure 8.3 provides the histograms of STIMULUS3 and STIMULUS7 received by the RANGER group and their corresponding behavior weights in the case of B1C1D1L1R25S2W8. From the figure, we can see that this group does not learn anything as time goes by. Thus, their motions are almost random. This can be described below as rule *RE* (see Figure 8.4):

$$\begin{aligned} &Rule\ RE: \\ &\qquad \forall i \in \texttt{WILD},\ \forall f_i \in \mathcal{S} \\ &\qquad\qquad \Downarrow \\ &P_i^{t+1} = \texttt{action}(\bowtie B_k,\ d_0,\ P_i^t), \end{aligned} \tag{8.1}$$

where $\bowtie B_k$ means a non-deterministic behavior B_k, and $\mathcal{S}$ is the set of all stimuli.

Figures 8.5(a), (b), and (c) present the macroscopic spatial distance between the RANGER robots, the distance between the RANGER and the WILD robots, and the selected sectors at each step, respectively.

Figures 8.6 and 8.7 show the resulting spatial distributions of the group robots in the above-mentioned two cases, at three different steps, respectively.

2. **Case B1C1D3L1R25S1W8 (short-range) vs. Case B1C1D3L1R25S2W8 (long-range)**: The difference between the present two cases and the previously mentioned two cases is that here the distribution of the WILD is relatively centralized. Figures 8.8 and 8.9 show the spatial distributions of the group robots at three corresponding steps in these two cases, respectively. We can note that the results are similar to those in the previous two cases.

   In the case of long-range sensing, it is observed that the system is unstable and that the group robots do not react to the stimuli they encounter. This is primarily because when a robot moves with an arbitrary small step in an arbitrary direction, other robots with a long-range sensing capability may not be able to detect such changes and hence fail to learn any useful behaviors. On the other hand, robots with short-range sensors do not have this problem.

8.2 Initial Spatial Distribution

In the experiments, we considered two kinds of initial distributions: one consists of randomly centralized WILD robots with randomly distributed RANGER robots, and another consists of randomly decentralized WILD robots with randomly distributed RANGER robots.

1. **Case B1C1D1L1R25S1W8 (decentralized) vs. Case B1C1D3L1R25S1W8 (centralized)**: The two cases are the same except that they consider different initial distributions of the WILD robots. The goal of the experiment is to examine how the RANGER robots learn to find and surround the WILD. In these two different cases, the RANGER group learns and acquires different kinds of experience. In the first case, because the distribution of the WILD robots is decentralized (to some extent, the distance between two robots is greater than their sensors can reach), the RANGER robots encounter fewer stimuli; and the stimuli they encounter trigger behaviors *following* and *pursuing*. That is to say, the results of learning in the case of B1C1D1L1R25S1W8 are useful for following and pursuing a WILD robot. Figure 8.10 shows the histograms of two stimuli and their corresponding behavior weights. From them, we can find that almost all RANGER robots over time encounter such stimuli. Furthermore, after 100 steps, almost all 25 RANGER robots encounter the unique STIMULUS2 for the remaining steps. The second and the third sectors are identified as the optimal directions for achieving their goal. That is why we observe that the RANGER robots follow the found WILD robots, as shown in Figure 8.11(a). The phenomenon of surrounding can also be found in Figure 8.11(a).

   In the case of B1C1D3L1R25S1W8, the RANGER robots encounter 14 different stimuli. The most significant ones are STIMULUS1, STIMULUS2, STIMULUS3, STIMULUS8, STIMULUS4, and STIMULUS7. Figure 8.12

presents the learning results of the RANGER group in this case. All of them have induced the same rules as *RA1*, *RB1*, *RB2*, *RB3*, *RB4*, and *RB5*. From them we conclude that the RANGER will adopt the strategy of surrounding whenever it encounters the WILD. Figure 8.11(b) shows the surrounding tendency in the RANGER robots.

2. **Case B1C1D1L2R25S1W8 (decentralized) vs. Case B1C1D3L2R25S1W8 (centralized)**: The difference between these two cases and the above two cases is that here the second behavior selection mechanism is adopted. From the results of Figures 8.13 and 8.14, we can readily arrive at the same conclusion as above.

8.3 Inverted Sigmoid Function

During group learning, after the evaluation of selected behaviors, behavior weights will be modified according to the observed effects. In doing so, an inverted sigmoid function is applied to the weights before normalizing behavior weight vectors. Our experiments have shown that if the inverted sigmoid function term is not used in weight modification, the weight of any well-performed behavior with respect to the current environment is increased quickly to 1, and the weight of an unsuitable behavior is tuned down quickly to 0. In this case, the system is immediately stabilized in a locally optimal state; all 0-weight behaviors will not be selected again. On the other hand, the 1-weight behavior will always be selected if robots encounter the same stimulus, even though this behavior may not be the best one.

Having introduced the inverted sigmoid function term, the behavior weights may approach but will never reach 1 or 0. Thus, every behavior will have a chance to be considered.

8.4 Behavior Selection Mechanism

Two mechanisms for behavior selection have been described in the preceding chapters. One is to select a behavior based on a probability distribution that corresponds to a behavior weight vector. The other is to choose the behavior that has the highest corresponding weight in the behavior vector. Through experimental validations, we have found that the maximum weight-based selection mechanism is less robust and more harsh than the former. The phenomenon can be readily noted by comparing the paired subfigures in Figures 8.15, 8.16, 8.17, and 8.18. From the macroscopic point of view, the cases with the maximum weight-based selection mechanism become stabilized faster than the cases with the former mechanism. In other words, the WILD in the cases with the latter mechanism creates a stronger attractive force to the RANGER, as may be noted by comparing the paired subfigures in Figures 8.19 and 8.20.

8.5 Motion Mechanism

In the preceding chapters, we have presented four different motion mechanisms for the WILD group. In our experiments, we are interested in knowing how the RANGER can become stabilized if the WILD uses a special mechanism.

First, if the positions of the WILD robots are fixed, the dynamic properties of the system completely depend on the learning and performance by the RANGER robots. In this case, the WILD cannot induce any complex local constraints to the RANGER. Thus, the number of stimuli encountered is quite small. As a result, it is quite easy for the RANGER group to learn and to reach a stable state. Now, if we let the WILD move with a group learning capability, the dynamic properties of the system in this case become more complex than the above case. However, because of the concurrent learning in the WILD, the system can still quickly reach a stable state, similar to the previous case. Next, we let the motion of the WILD group become completely random. Thus, the RANGER robots encounter more stimuli. This makes the RANGER learning more difficult than the above cases. The system eventually reaches a dynamically stable state. Figures 8.21(a), (b), and (c) show the processes of learning how to react to a key stimulus, STIMULUS1, in the cases of B1C1D1L1R25S1W8, B2C1D1L1R25S1W8, and B3C1D1L1R25S1W8, respectively. Figures 8.22(a), (b), and (c) show the processes of learning how to react to another key stimulus, STIMULUS2, in the three cases, respectively.

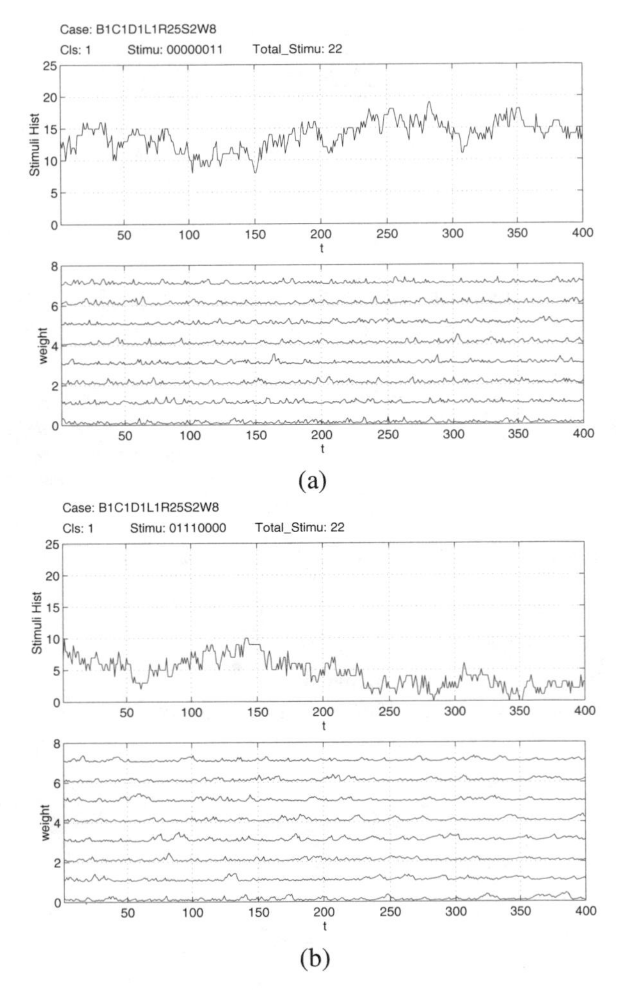

(b)

FIGURE 8.3. The histograms of (a) STIMULUS3 and (b) STIMULUS7 received by the RANGER group and their corresponding behavior weights in the case of B1C1D1L1R25S2W8.

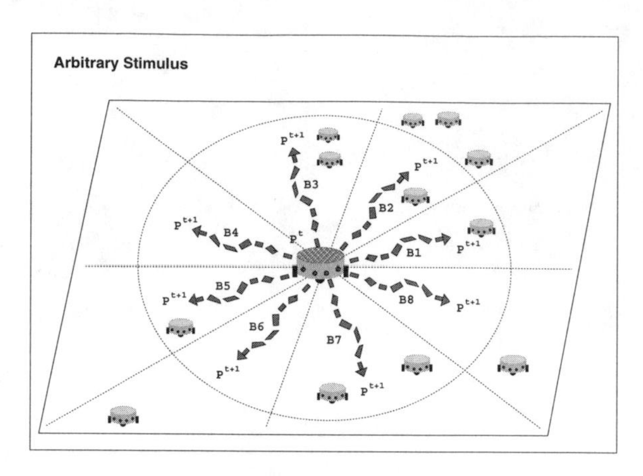

FIGURE 8.4. An illustration of rule *RD3*.

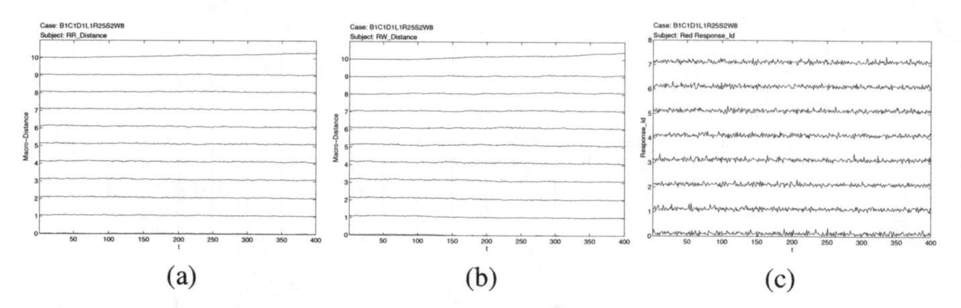

(a) (b) (c)

FIGURE 8.5. The distributions of the spatial distances (a) between two RANGER robots and (b) between a RANGER and a WILD robot. (c) Behaviors selected by the RANGER robots.

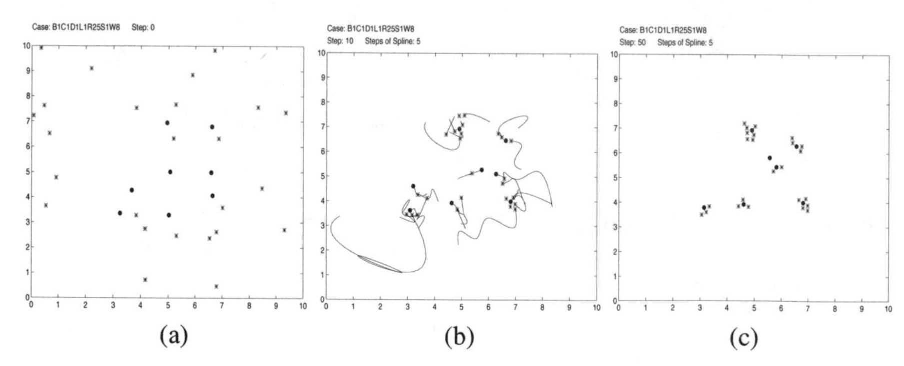

(a) (b) (c)

FIGURE 8.6. (a) The original distribution of the robots in the case of B1C1D1L1R25S1W8. (b) The spatial distribution of the robots at step 10. (c) The spatial distribution of the robots at step 50.

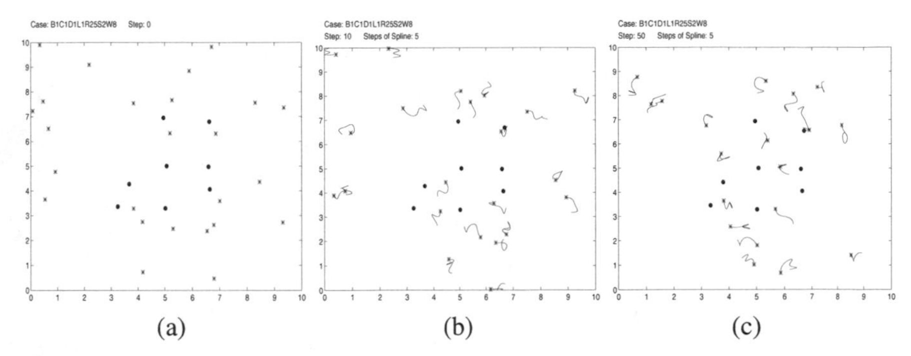

FIGURE 8.7. (a) The original distribution of the robots in the case of B1C1D1L1R25S2W8. (b) The spatial distribution of the robots at step 10. (c) The spatial distribution of the robots at step 50. Note that some RANGER robots have moved outside the environment.

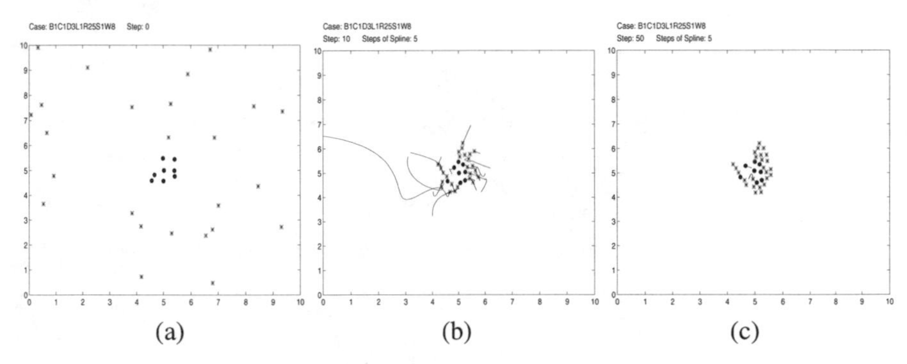

FIGURE 8.8. (a) The original distribution of the robots in the case of B1C1D3L1R25S1W8. (b) The spatial distribution of the robots at step 10. (c) The spatial distribution of the robots at step 50.

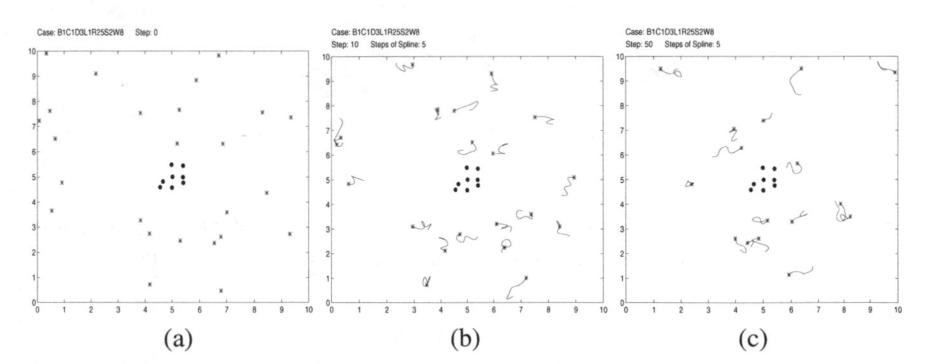

FIGURE 8.9. (a) The original distribution of the robots in the case of B1C1D3L1R25S2W8. Note that it is the same as the one in Figure 8.8(a). (b) The spatial distribution of the robots at step 10. (c) The spatial distribution of the robots at step 50. Note that some RANGER robots have moved outside the environment.

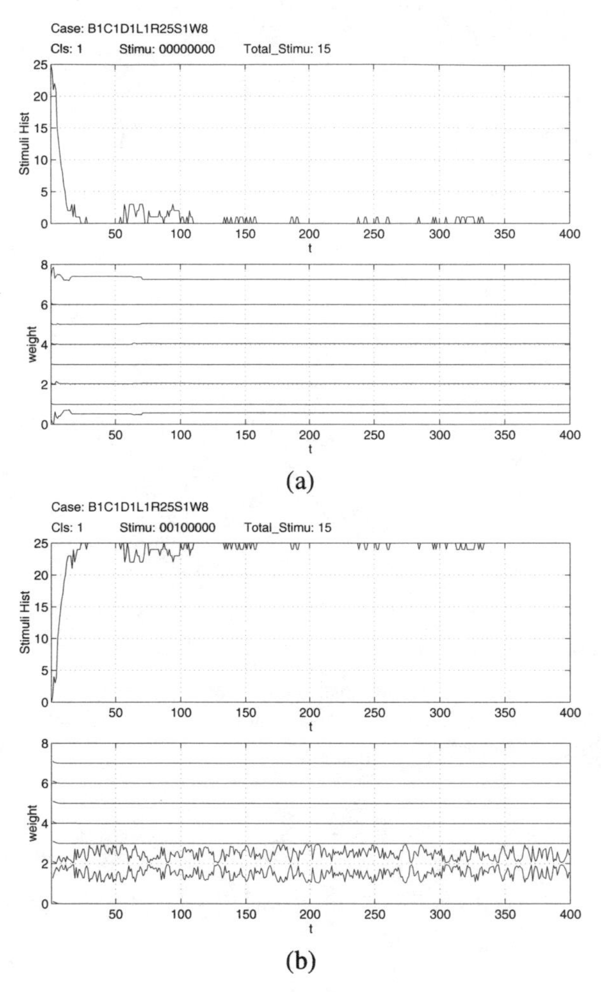

FIGURE 8.10. The histograms of (a) STIMULUS1 and (b) STIMULUS2 received by the RANGER group and their corresponding behavior weights in the case of B1C1D1L1R25S1W8.

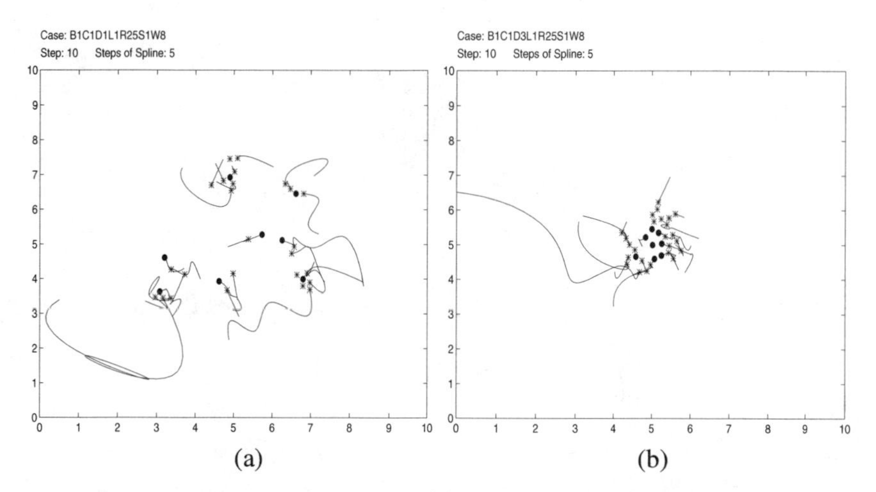

FIGURE 8.11. The spatial distributions of the robots at step 10 in the cases of (a) B1C1D1L1R25S1W8 and (b) B1C1D3L1R25S1W8, respectively.

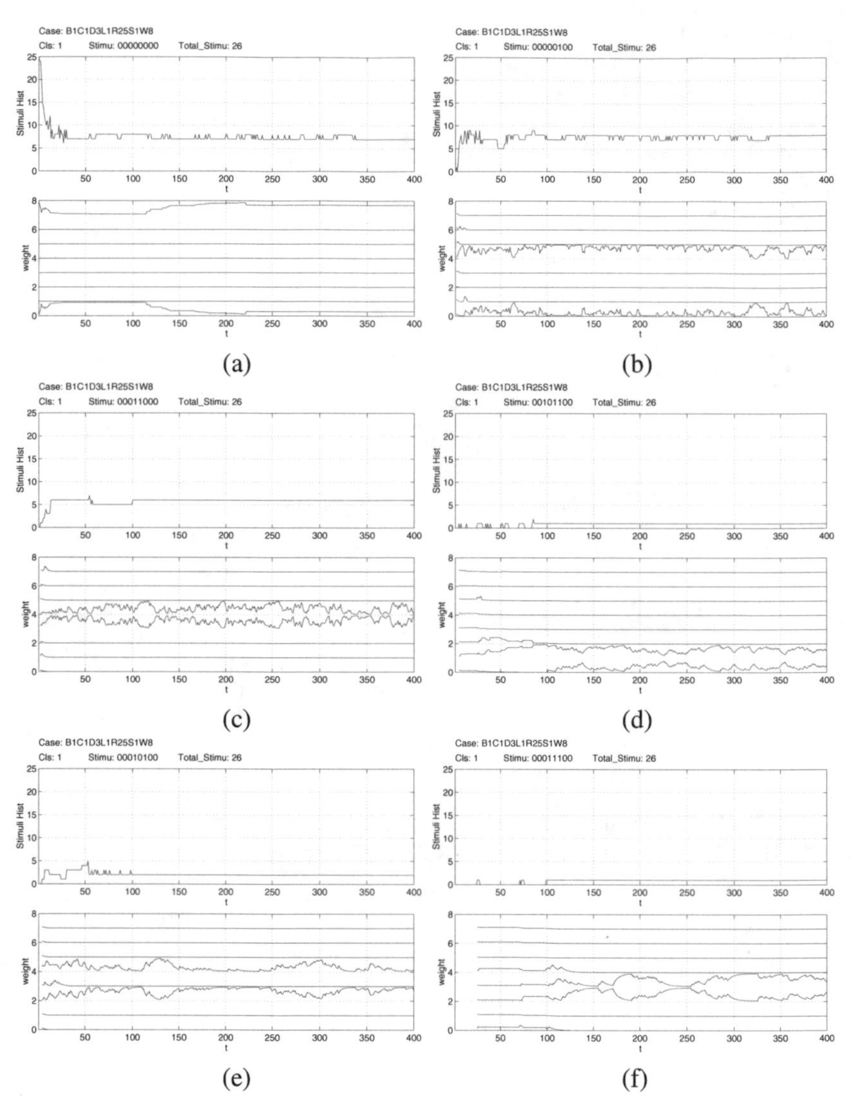

FIGURE 8.12. The histograms and corresponding behaviors with respect to (a) STIMULUS1, (b) STIMULUS2, (c) STIMULUS3, (d) STIMULUS8, (e) STIMULUS4, and (f) STIMULUS7 in the RANGER group in the case of B1C1D3L1R25S1W8.

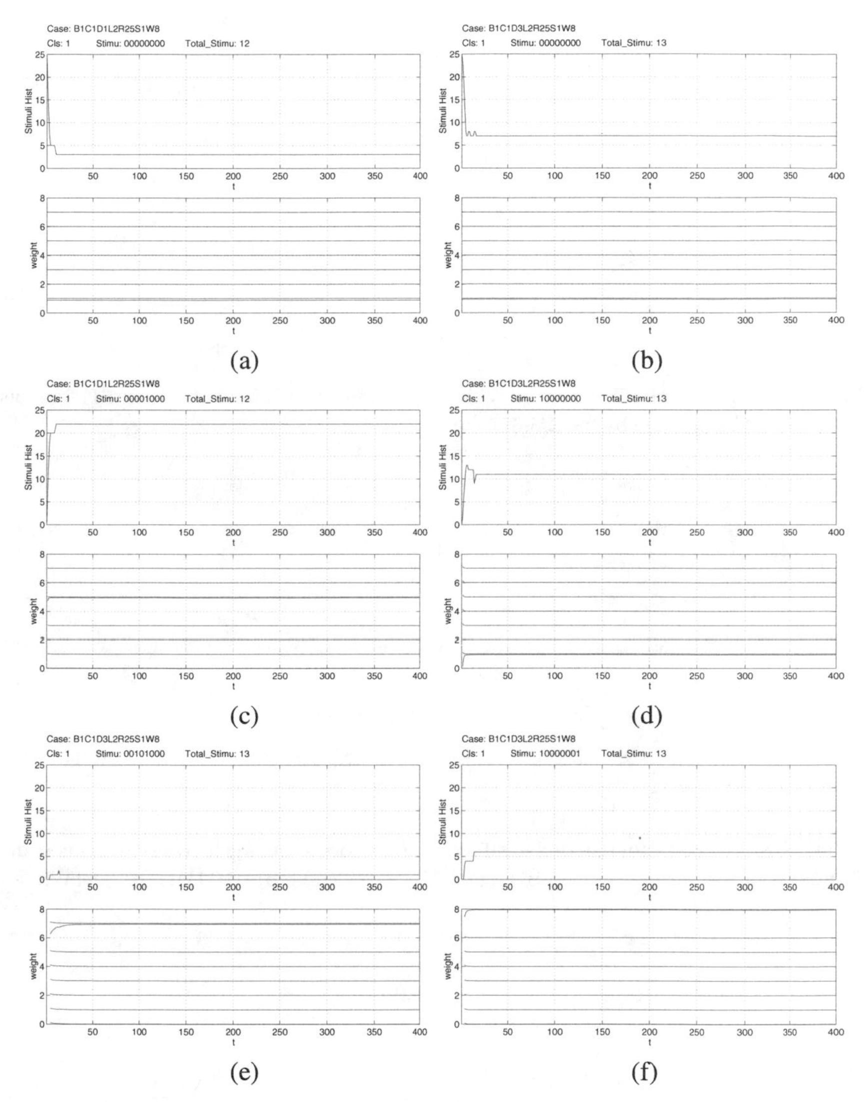

FIGURE 8.13. The histograms and corresponding behaviors of the RANGER group with respect to (a) STIMULUS1 in the case of B1C1D1L2R25S1W8; (b) STIMULUS1 in the case of B1C1D3L2R25S1W8; (c) STIMULUS2 in the case of B1C1D1L2R25S1W8; (d) STIMULUS2 in the case of B1C1D3L2R25S1W8; (e) STIMULUS4 in the case of B1C1D3L2R25S1W8; and (f) STIMULUS3 in the case of B1C1D3L2R25S1W8.

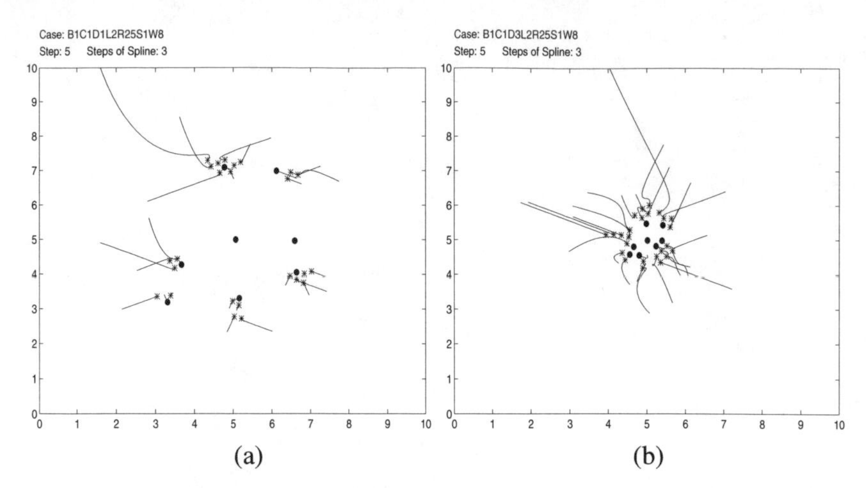

FIGURE 8.14. All RANGER robots move directly toward the WILD robots. Some WILD robots are followed or pursued by the RANGER robots.

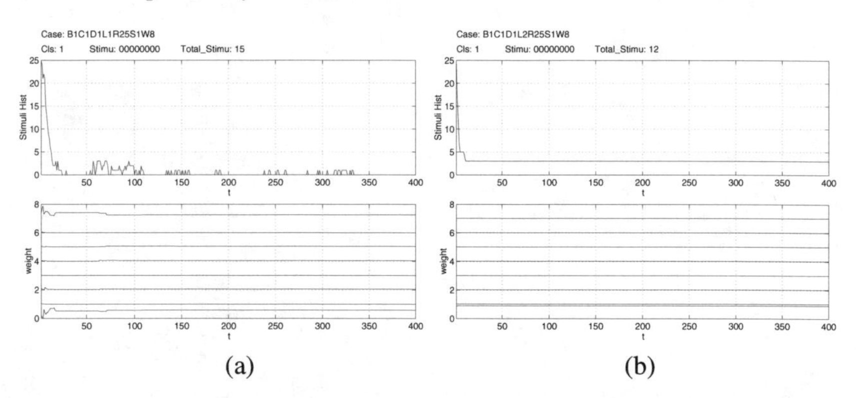

FIGURE 8.15. The histogram of STIMULUS1 and its corresponding behavior weights in the RANGER group in the cases of (a) B1C1D1L1R25S1W8 and (b) B1C1D1L2R25S1W8.

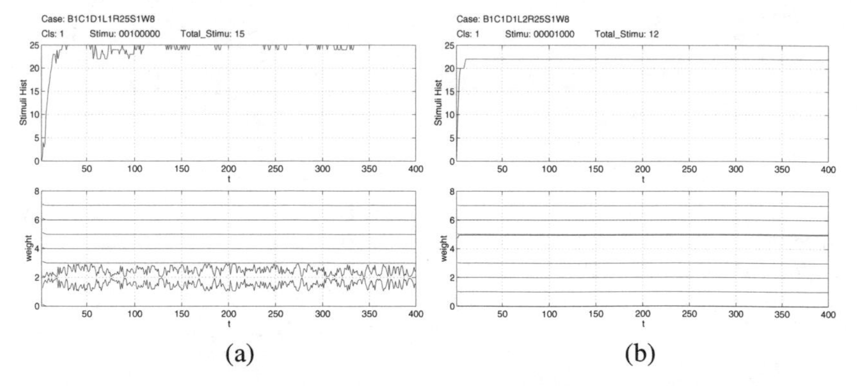

FIGURE 8.16. The histogram of STIMULUS2 and its corresponding behavior weights in the RANGER group in the cases of (a) B1C1D1L1R25S1W8 and (b) B1C1D1L2R25S1W8.

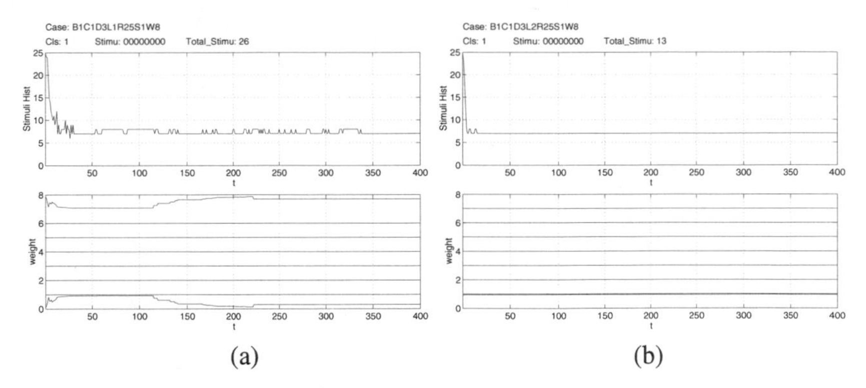

(a) (b)

FIGURE 8.17. The histogram of STIMULUS1 and its corresponding behavior weights in the RANGER group in the cases of (a) B1C1D3L1R25S1W8 and (b) B1C1D3L2R25S1W8.

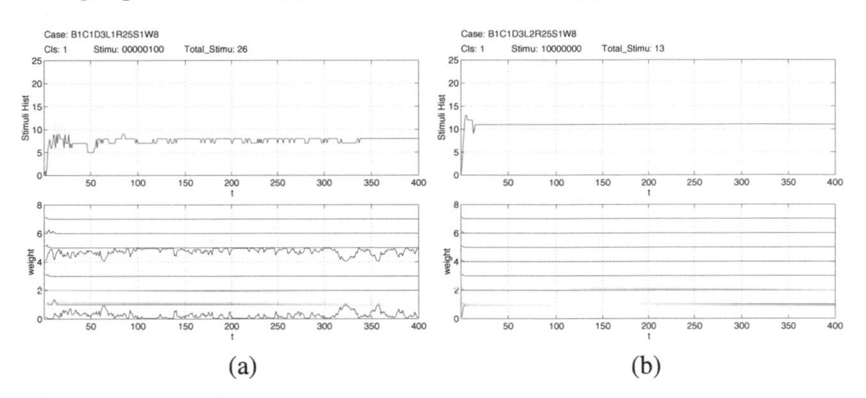

(a) (b)

FIGURE 8.18. The histogram of STIMULUS2 and its corresponding behavior weights in the RANGER group in the cases of (a) B1C1D3L1R25S1W8 and (b) B1C1D3L2R25S1W8.

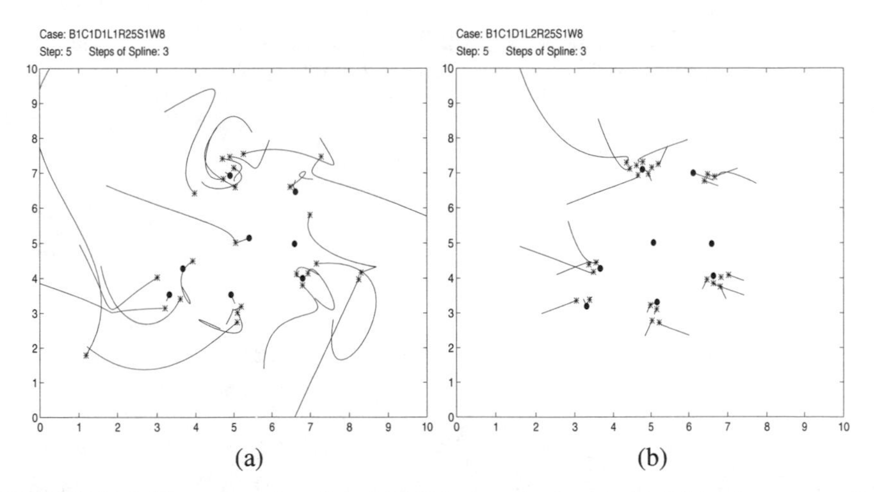

(a) (b)

FIGURE 8.19. The trajectories of the RANGER robots in chasing the WILD robots at step 5, in the cases of (a) B1C1D1L1R25S1W8 and (b) B1C1D1L2R25S1W8.

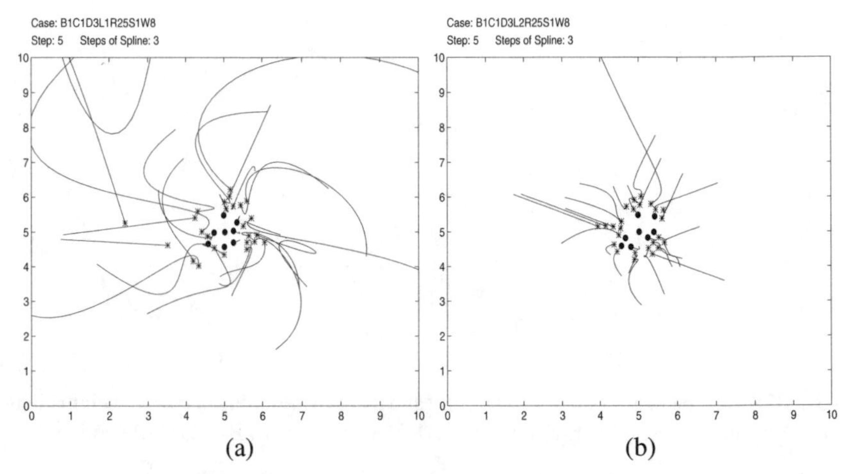

FIGURE 8.20. The trajectories of the RANGER robots in chasing the WILD robots at step 5, in the cases of (a) B1C1D3L1R25S1W8 and (b) B1C1D3L2R25S1W8.

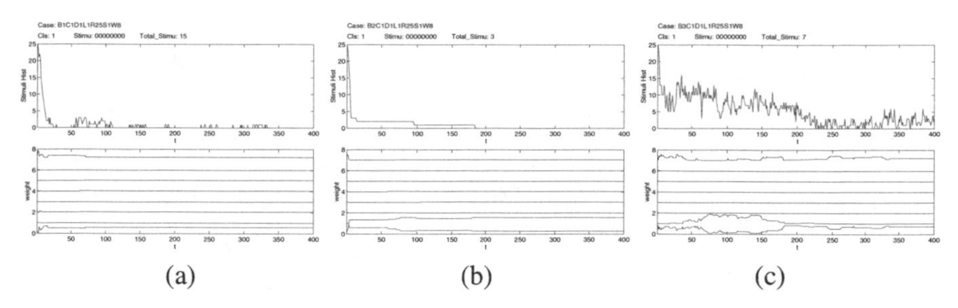

FIGURE 8.21. The histogram and corresponding behavior weights for STIMULUS1 in the RANGER group in the cases of (a) B1C1D1L1R25S1W8, (b) B2C1D1L1R25S1W8, and (c) B3C1D1L1R25S1W8.

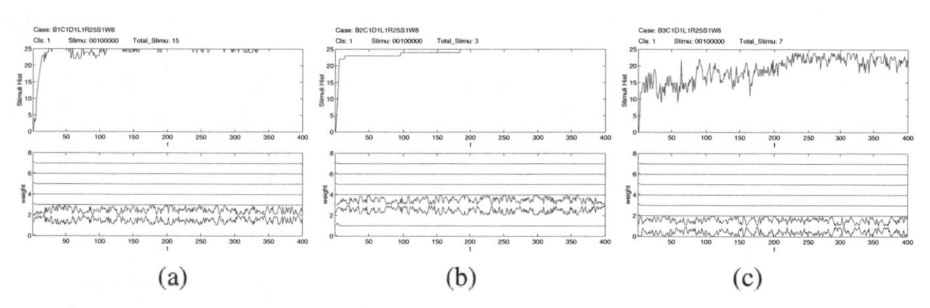

FIGURE 8.22. The histogram and corresponding behavior weights for STIMULUS2 in the RANGER group in the cases of (a) B1C1D1L1R25S1W8, (b) B2C1D1L1R25S1W8, and (c) B3C1D1L1R25S1W8.

8.6 Emerging a Periodic Motion

As observed from our experiments, long-range sensors can sometimes induce periodic motions in the RANGER robots.

From Figures 8.23(a) and (b), we can note that there are three RANGER robots in the upper part of the figures and three in the lower part moving periodically. Their periodic motions are induced from the periodic stimuli, as shown in Figure 8.24. When a RANGER robot encounters STIMULUS3 (00000011), it selects the first sector to move into. Then, its stimulus will change to 01110000 and it will choose behavior 1 to react. As a result, it will go back to an environment from which the received stimulus is 00000011 again, as shown in Figure 8.23(c). Thus, it moves periodically.

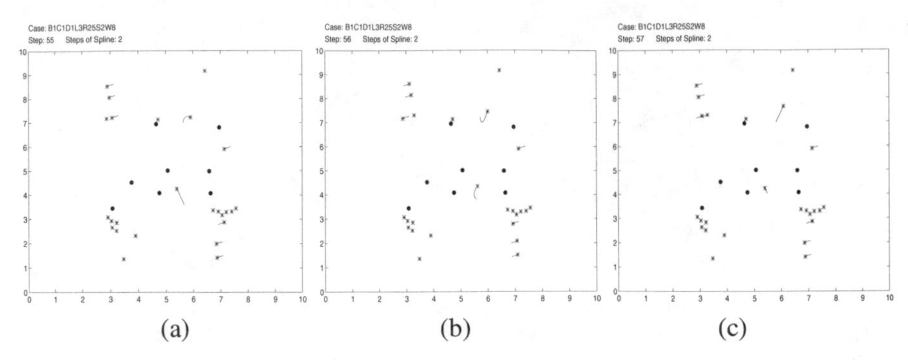

FIGURE 8.23. Some RANGER robots move periodcally, as shown at steps (a) 55, (b) 56, and (c) 57.

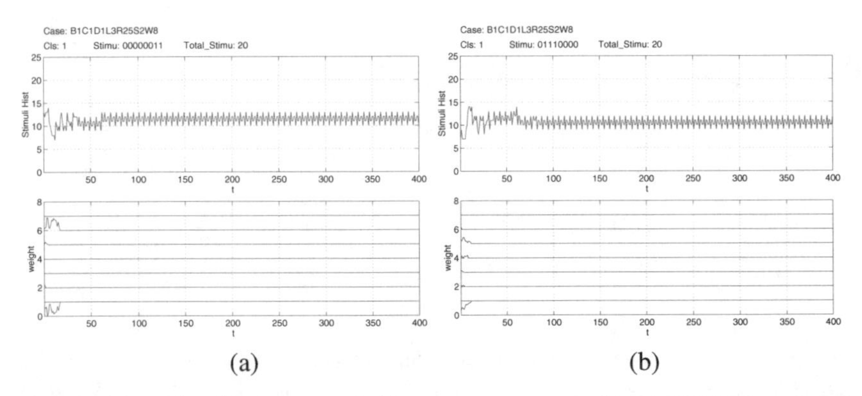

FIGURE 8.24. The histograms of (a) `STIMULUS3` and (b) `STIMULUS7` and their corresponding behavior weights in the `RANGER` group in the case of B1C1D1L3R25S2W8.

8.7 Macro-Stable but Micro-Unstable Properties

In some cases, although a system has reached a globally stable state, some individual robots may still be unstable. Figure 8.25(a) shows a stable distribution of the system in the case of B1C1D4L1R25S1W8. Figure 8.26 presents the learning processes of the `WILD` group encountering `STIMULUS2` and `STIMULUS4`, respectively. From this figure, we can realize that the weights are unstable. In other words, the `WILD` robots cannot decide what to do when they encounter such stimuli. The cause of such a phenomenon is that the `WILD` robots have been surrounded by some `RANGER` robots or surrounded by other `WILD` robots. No matter which behavior it selects, it cannot change its current state of being surrounded. Therefore, it tries out every behavior but without much success. In this case, the `WILD` robots also encounter other stimuli after the system has become stable, such as `STIMULUS3`, `STIMULUS8`, and `STIMULUS13`. Figure 8.25(b) shows the circumstance of encountering `STIMULUS8`. With these stimuli, the `WILD` group is close to the `RANGER`. Thus, their slight motions can change their stimuli as well as local constraints from which the robots can gradually learn something. The cause of this phenomenon is related to the parameters of the robot sensors. It may be possible that after the sensing radius is changed, they can sense useful information and detect local constraints.

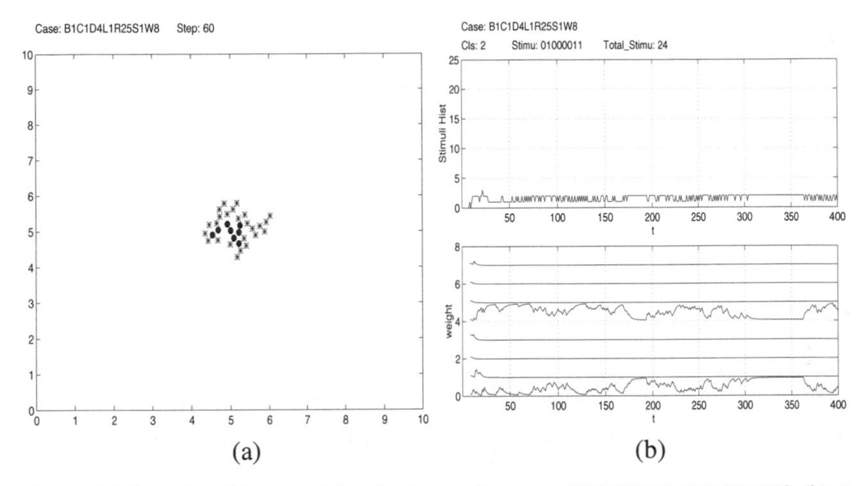

FIGURE 8.25. (a) A stable spatial distribution in the case of B1C1D4L1R25S1W8. (b) An example of learning with respect to STIMULUS8 in this case.

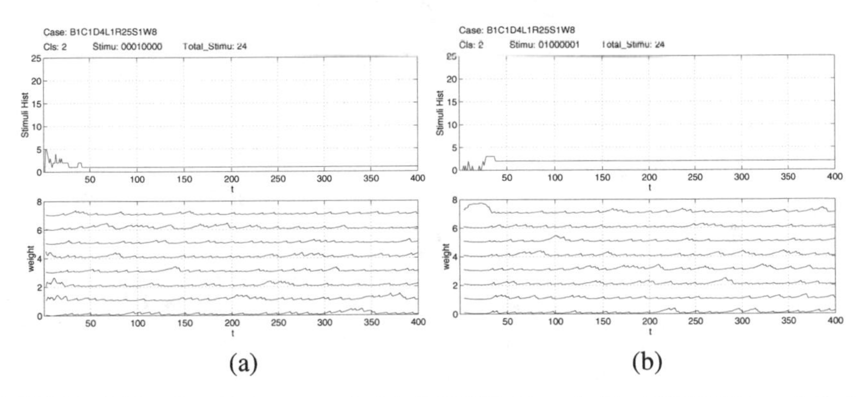

FIGURE 8.26. The histograms of (a) STIMULUS2 and (b) STIMULUS4 and their corresponding behavior weights in the WILD group in the case of B1C1D4L1R25S1W8.

8.8 Dominant Behavior

Most of the learning results in the cases of B*C1D*L1R25S1W8 have shown that the main reaction to a certain stimulus always consists of two behaviors. We can observe this from some of the previous examples, such as Figure 8.27. This demonstrates that neither of the two behaviors is dominantly better than the other.

After a robot selects, performs, and evaluates one of them, it may still switch to another one later. The cause of this phenomenon is that the resolution of sensors with respect to the environment is not high enough.

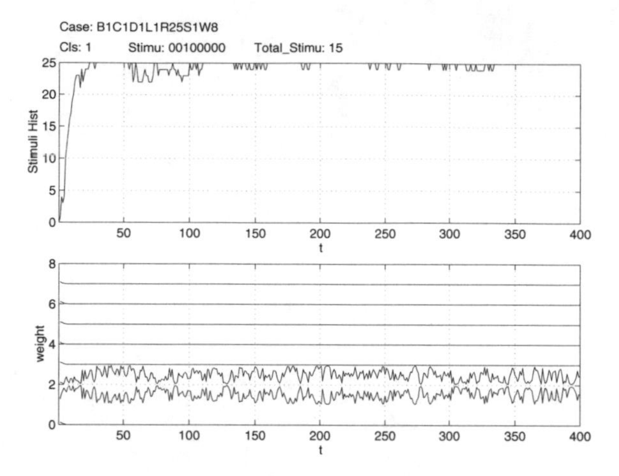

FIGURE 8.27. The histogram of STIMULUS2 and its corresponding behavior weights in the RANGER group in the case of B1C1D1L1R25S1W8.

9

Evolutionary Multi-Agent Reinforcement Learning

The radical of one century is the conservative of the next. The radical invents the views. When he has worn them out, the conservative adopts them.

Mark Twain

In a system of decentralized autonomous robots, each individual robot can have its own primitive behaviors, such as avoidance, following, aggregation, dispersion, homing, and wandering. These behaviors are precisely defined through an array of behavior parameters (for controlling sensing and stimulus extraction capabilities, reactive motion strategies, and reinforcement strengths, etc.). The robots belonging to one group may share some or all of their behavior characteristics.

In this chapter, we will discuss the problem of acquiring emergent behavior among several decentralized mobile robotic agents and address the issue of interrelationships between the autonomy of individual robots and their group evolution. Specifically, we will describe how to enable a group of robots to determine their behavior characteristic parameters in order to achieve a globally optimal performance.

We will present a genetically controlled means for a group of autonomous robots (RANGER) to collectively achieve such tasks as surrounding a group of WILD robots. Toward this end, we will demonstrate how a genetic algorithm can be applied to evolve some globally optimal group behavior among the robots, by way of selecting appropriate sensory capability and behavior learning or selection capability[1]. At the same time, we will identify the most effective task environment, such as the specific targets of certain spatial distribution and motion characteristics.

9.1 Robot Group Example

In our present work, we will study both the local and the global behaviors of a group of 25 simulated RANGER robots as denoted by the $*$ symbols in the spatial maps of Figures 9.1 and 9.2. The individual mobile robot of this group can receive its time-varying stimulus that reflects the changes in its environment, that is, the change in the number of WILD robots within its neighboring region before and after the execution of a selected behavior. The WILD robots are marked with the $\bullet$ symbols in the maps. Without loss of generality, the task of the autonomous robots that we will concentrate on here is to surround the group of eight WILD robots placed within a 10×10 area.

9.1.1 *Target Spatial Distributions*

The spatial characteristics of the WILD robots can be described in terms of two types of distributions: (1) random distribution over a relatively large region (radius of 4 units), which may be regarded as a decentralized distribution, and (2) random distribution over a small region (radius of 1 unit), which may be regarded as a centralized distribution. Figures 9.1 and 9.2 provide two examples of such spatial distributions, respectively.

9.1.2 *Target Motion Characteristics*

In addition to the spatial distributions, the WILD robots may further be described by their motion characteristics. Here we consider the WILD robots of three distinct motion characteristics; namely, (1) learning to move away from the chasing robots, (2) stationary, and (3) random motion.

---

[1]We assume that all the robots being considered here are homogeneous in their functionality as well as their behavior learning capability and can communicate among themselves in order to share their reactive behavior strategies.

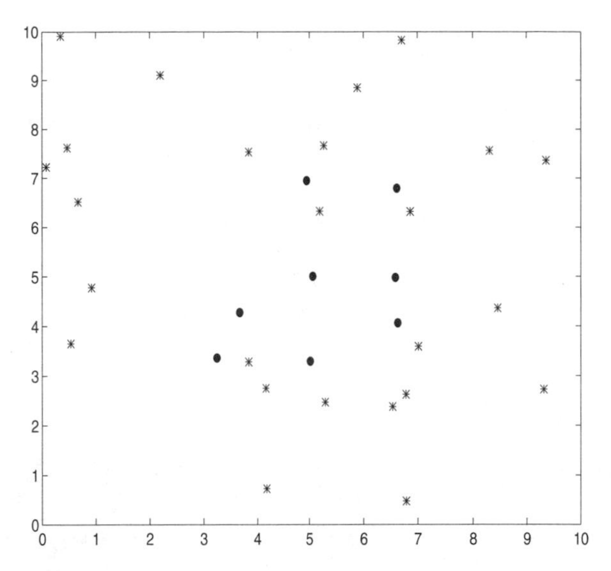

FIGURE 9.1. A typical example of decentralized WILD distribution in a robot task environment (©1998 IEEE).

9.1.3 *Behavior Learning Mechanism*

Each robot has a built-in reinforcement learning (RL) mechanism. Suppose that the robot selects behavior B_k based on the probability distribution of its behavior weight vector, while encountering stimulus $\int_k$. After the execution of the behavior, the corresponding behavior weights vector $W^t_{\int_k}$ are updated according to the following function:

$$W^{t+1}_{\int_k} = \texttt{normal}(\texttt{smooth}(W^t_{\int_k} + \Delta W)), \tag{9.1}$$

where ΔW is a weight increment vector whose j component is defined as follows:

$$\Delta w_j = \begin{cases} \delta \mid_{E(B_k)}, & \text{if } j = k, \\ 0, & \text{if } j \neq k. \end{cases} \tag{9.2}$$

The $E(B_k)$ function evaluates the result of behavior B_k, in terms of change in the number of WILD robots inside the neighboring region of the robot, and $\delta \in [-1, 1]$. Operator smooth performs a post-processing operation to smooth the weights by scaling down the relatively larger weights and scaling up the smaller ones. Operator normal normalizes the weight vector.

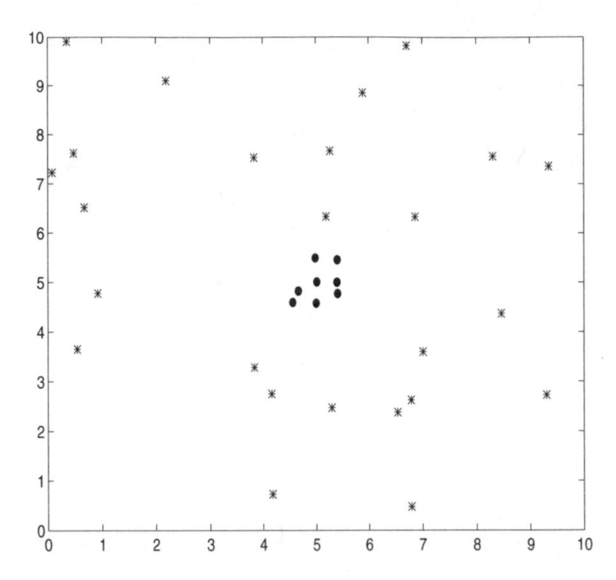

FIGURE 9.2. A typical example of centralized `WILD` distribution in a robot task environment (©1998 IEEE).

9.2 Evolving Group Motion Strategies

The preceding chapters have described the sensory and behavioral capabilities of an autonomous robot and the characteristics of its task environment. What follows will address the central issue of this chapter – that is, given a specific spatial or temporal optimality criterion for a group of autonomous mobile robots that collectively perform a predefined task (such as surrounding a group of `WILD` robots), find the most suitable sensory and behavioral requirements in the individual robots, and determine the most effective task configuration, which would enable the robots to exhibit the optimal group performance.

Specifically, we will apply a genetic algorithm to evolve a near-optimal solution. The algorithm employs a set of operators to gradually produce better *generations* of solutions. Here the fitness function of each member in a population is computed by measuring how well it satisfies a given optimality criterion. The best members of the population are rewarded in their fitness values and hence are selected, while poorly performing members are removed. In this way, the population, and thus some of the individuals, can quickly converge to some near-optimal solution(s) to the group behavior optimization problem.

9.2.1 Chromosome Representation

A 7-bit chromosome will be used in the genetic algorithm-based group evolution, which encodes the specific characteristics as mentioned in Section 9.1. Figure 9.3

shows the definition of the chromosome that expresses up to 96 distinct conditions. The semantics of the individual bits are described in Table 9.1.

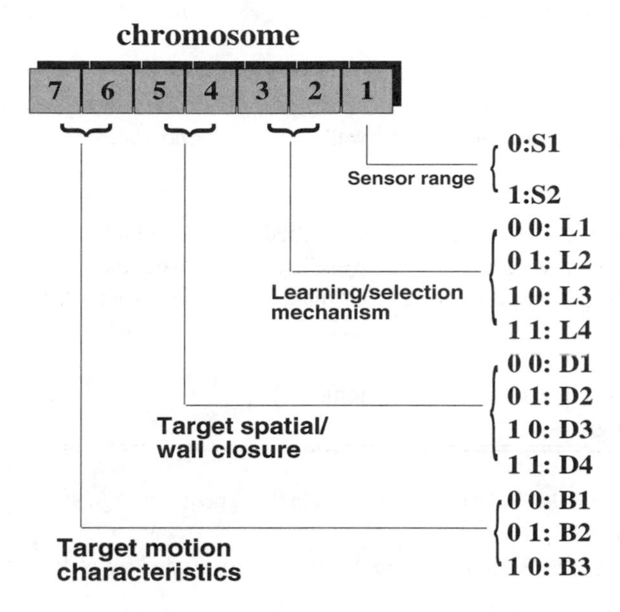

FIGURE 9.3. Definition of the chromosome (©1998 IEEE).

9.2.2 *Fitness Functions*

In our experiments, we define an optimal group performance in terms of the fitness functions of individual robots. In this respect, four types of fitness functions will be investigated, which are defined as follows:

1. `CONTACT`: Each `RANGER` robot can find at least one `WILD` robot, that is:

$$s_1 = \frac{T_1 - T_0}{\sum_{t=T_0}^{T_1} D^t}, \tag{9.3}$$

where

$$D^t = \frac{1}{M_R} \sum_{i=1}^{M_R} \mathtt{min}\{\mathtt{dist}(P_i^t, P_j^t) | \, j = 1, \ldots, M_W\}. \tag{9.4}$$

This fitness function calculates the average distance between an autonomous robot i and the closest `WILD` robot j, at time t. M_R and M_W denote the total numbers of `RANGER` robots and `WILD` robots, respectively. Function `dist` computes the distance between `RANGER` robot i and `WILD` robot j.

	1	2	3	4
B: Target motion characteristic	learning to move away	stationary	random motion	
D: Target distribution, wall-closure	decentralized, no wall	decentralized, wall	centralized, no wall	centralized, wall
L: Learning, selection mechanism	smoothed weights, by probability	smoothed weights, by maximum weight	original weights, by probability	original weights, by maximum weight
S: Sensor range	short	long		

TABLE 9.1. Selectable conditions in the experiments (©1998 IEEE).

2. `SURROUND`: Each `WILD` robot will be surrounded by at least three robots, that is:

$$s_2 = \frac{1}{\sum_{t=T_0}^{T_1} (\bar{D}^3_{t,W})^2}, \tag{9.5}$$

where

$$\bar{D}^k_{t,W} = \frac{1}{M_W} \sum_{i=1}^{M_W} \left\{ \frac{1}{k} \sum^{k} \Psi_k \{\mathtt{dist}(P_i^t, P_j^t) \,|\, j = 1, \ldots, M_R\} \right\}. \tag{9.6}$$

This fitness function computes the average distance between a `WILD` robot i, and the nearest k `RANGER` robots at time t. Function Ψ_k computes the k minimum distances.

3. `CHASE`: Each `RANGER` robot moves directly toward the `WILD`, that is:

$$s_3 = \frac{1}{\sqrt{\sum_{k=1}^{M_R} \sum_{t=T_0}^{T_1} d(t,k)}}, \tag{9.7}$$

where $d(t, k)$ is the movement step of robot k at time t.

4. `FOCUS&CONTACT`: Each `RANGER` robot focuses on learning some specific reactive behaviors and attains the goal of finding at least one `WILD` robot, that is:

$$s_4 = s_1 \cdot \mathtt{std}\left(\sum_{t=T_0}^{T_1} w_{f^t_{max}}\right), \tag{9.8}$$

where f^t_{max} corresponds to the most frequently encountered stimulus at time t. $\sum_{t=T_0}^{T_1} w_{f^t_{max}}$ sums up the behavior weights for respective stimuli. Function `std` computes the standard deviation of the weight sum.

Here it may be noted that this fitness function is concerned not only with the `CONTACT` behavioral requirement but also the requirement for high-efficiency learning based on a limited number of stimuli.

9.2.3 *The Algorithm*

The complete genetic algorithm as used in the group behavior evolution is given in Figure 9.4.

begin
  define fitness function s_i,
  define the maximum number of generations per step $\mathcal{G}$,
  define population size $\mathcal{P}$,
  define crossover probability p_c,
  define mutation probability p_m,
  modify *seed* of random number generator,
  create an initial population of $\mathcal{P}$ members,
  for $generation$: $1 \longrightarrow \mathcal{G}$ **do**
    evaluate each individual fitness s_i in current $generation$:
      for population: $1 \longrightarrow \mathcal{P}$ **do**
        modify sensory/behavioral conditions for the group,
        enable low-level robot learning,
        observe emergent group behavior,
        evaluate group performance,
      endfor
    copy the best individuals to the next $generation$,
    select other individuals based on their fitness values,
    use one-point crossover with probability p_c,
    mutate the individuals of $generation$ with probability p_m,
  endfor
end

FIGURE 9.4. The genetic algorithm for selecting robot sensory/behavioral characteristics as well as corresponding task configurations.

9.2.4 *Parameters in the Genetic Algorithm*

Parameter	L	$\mathcal{P}$	$\mathcal{G}$	p_c	p_m	T_0	T_1
Value	7	7	40	0.6	Eq. 9.9	1	20

TABLE 9.2. Parameters as used in the experiments (©1998 IEEE).

The specific parameters in our genetic algorithm are given in Table 9.2, where L denotes the bit length of a chromosome, and mutation probability p_m is defined as follows:

$$p_m = \begin{cases} 0.2, & \text{if } 1 \leq generation < 10, \\ 0.06, & \text{if } 10 \leq generation < 20, \\ 0.005, & \text{if } 20 \leq generation < 30, \\ 0.001, & \text{if } 30 \leq generation \leq 40. \end{cases} \tag{9.9}$$

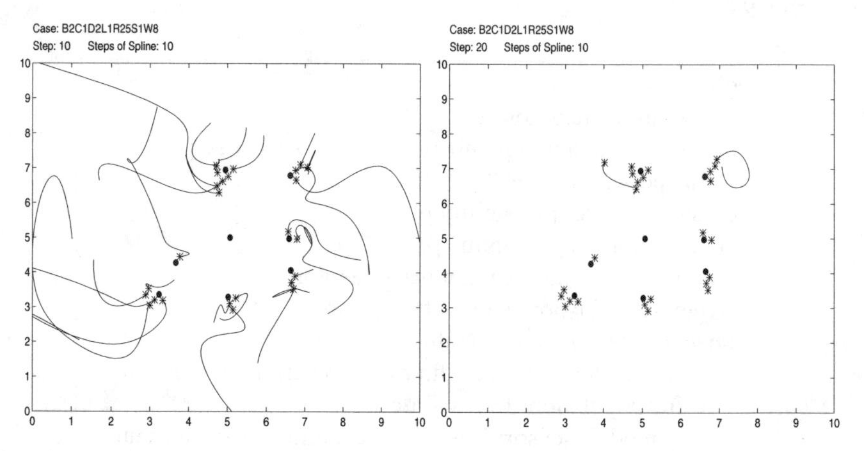

FIGURE 9.5. The robot motion trajectories in the case of B2C1D2L1R25S1W8 (©1998 IEEE).

9.3 Examples

As examples of the genetic algorithm based evolution, Figures 9.5 to 9.8 present the trajectories of motion in surrounding a group of WILD robots by a group of learning robots, which exhibit the optimal group behavior as defined using the four fitness functions, respectively. In the figures, the left-hand side shows the first 10 steps of robot motion, and the right-hand side corresponds to the next 10 steps. In order to gain a better insight into the process of the evolution, Figures 9.9 to 9.12 show the detailed changes in the four fitness functions over 40 generations, respectively.

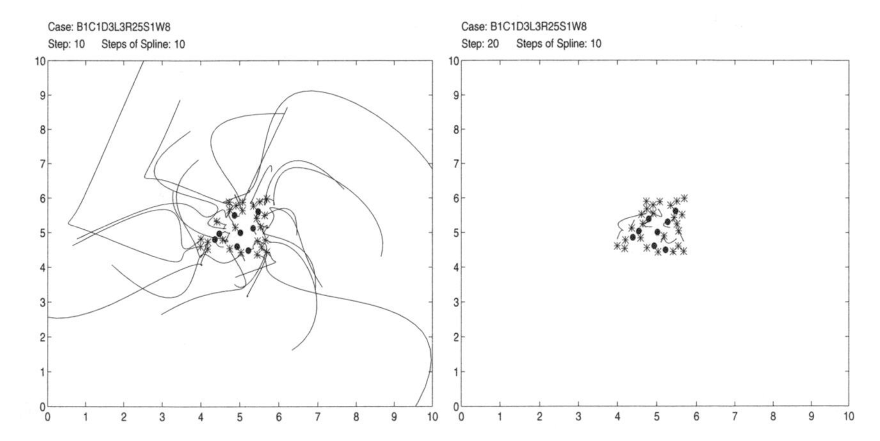

FIGURE 9.6. The robot motion trajectories in the case of B1C1D3L3R25S1W8 (©1998 IEEE).

From the above results, it can be readily noted that all the trajectories as well as the converged spatial locations of the robots after 20 movement steps meet the predefined optimality requirements quite well. In addition, we can make the following observations on the results of group behavior evolution:

1. It is very effective to perform the CONTACT group behavior in a task configuration where the WILD robots are stationary and in a decentralized spatial distribution. In such a case, the best way for an individual RANGER robot to select its movement from its behavior weight vector is based on the probability distribution after a smoothing operation.

2. Both SURROUND and CHASE are effective with moving WILD targets with centralized spatial distributions. In addition, in order to achieve the former group behavior, it requires each of the individual RANGER robots to select its own local behavior based on non-smoothed probability distribution, allowing a certain degree of randomized movements for the robots to evenly surround the WILD group. For the latter group behavior, however, it is most efficient for the individual robots to determine their movements by selecting the maximum weight behavior responses.

3. Generally speaking, the WILD robots of a centralized spatial distribution will have a stronger attraction. In Figures 9.6 and 9.7, all the RANGER robots settle around the WILD robots after 20 steps.

4. Unlike the CONTACT behavior, FOCUS&CONTACT is effective in the cases of moving targets with decentralized spatial distribution. This is because the moving WILD targets can readily improve the behavior learning in individual RANGER robots by offering a variety of stimuli for the robots to search and acquire their reactive behaviors accordingly. The target-

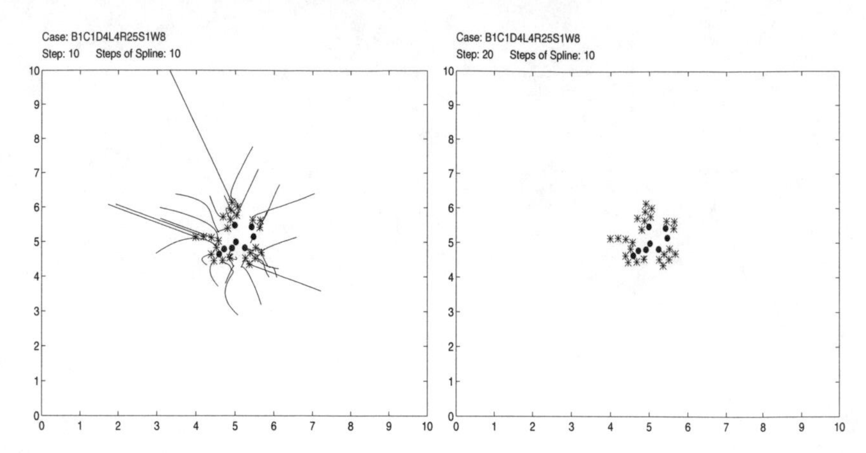

FIGURE 9.7. The robot motion trajectories in the case of B1C1D4L4R25S1W8 (©1998 IEEE).

following behavior, as can be observed from the trajectories of the robots, also shows that learning by the RANGER robots in such a case is quite robust.

9.4 Summary

This chapter has presented an evolutionary computation approach to selecting robot sensory, behavioral, and task configurations to allow for the emergence of optimal group behavior. The multi-agent approach can be regarded as an effective way of developing, predicting, and controlling the group behavior of autonomous robots in a distributed setting.

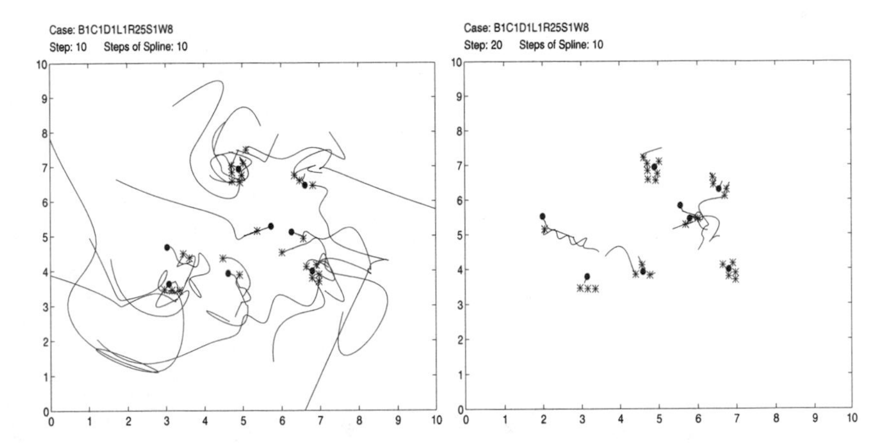

FIGURE 9.8. The robot motion trajectories in the case of B1C1D1L1R25S1W8 (©1998 IEEE).

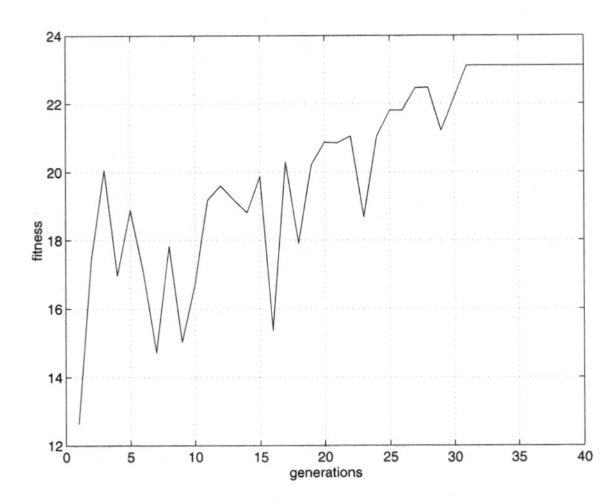

FIGURE 9.9. The change of s_1 (CONTACT) fitness over 40 generations (©1998 IEEE).

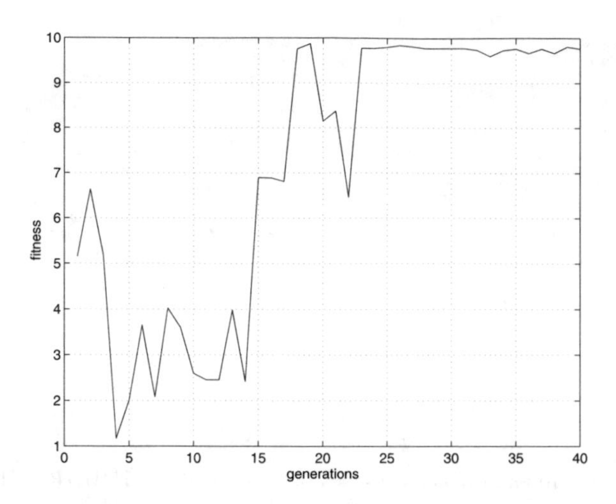

FIGURE 9.10. The change of s_2 (SURROUND) fitness over 40 generations (©1998 IEEE).

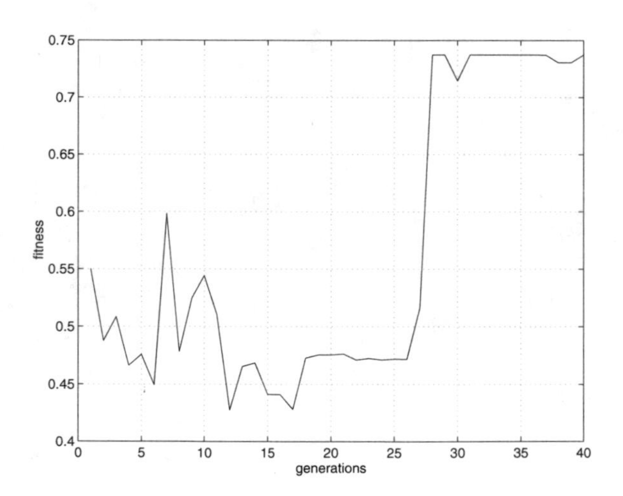

FIGURE 9.11. The change of s_3 (CHASE) fitness over 40 generations (©1998 IEEE).

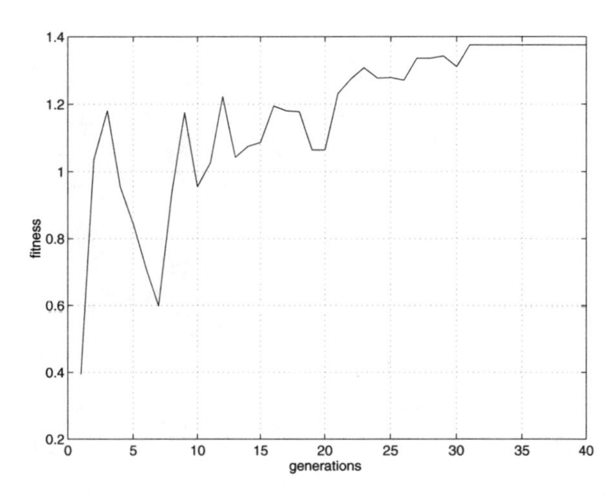

FIGURE 9.12. The change of s_4 (FOCUS&CONTACT) fitness over 40 generations (©1998 IEEE).

Part III

Case Studies in Adaptation

10
Coordinated Maneuvers in a Dual-Agent System

The real problem is not whether machines think but whether men do.

B. F. Skinner

Robotic agents adapt to their dynamically changing environments by performing coordinated maneuvers. Such maneuvers may be achieved either at the level of actuators in a single robot or at the level of robot groups.

In this chapter, we will describe a dual-agent system capable of evolving eye-body coordinated maneuvers in a sumo contest. The two agents rely on each other by either offering feedback information on the physical performance of a selected maneuver or giving advice on candidate maneuvers for an improvement over the previous performance. Central to this learning system is a *multi-phase genetic programming* approach that is aimed at enabling the player to gradually acquire sophisticated sumo maneuvers. As will be illustrated in the sumo learning experiments involving opponents of complex shapes and sizes, the developed multi-phase learning allows the development of specialized strategic maneuvers based on general ones and hence demonstrates the effectiveness of maneuver acquisition.

In this chapter, we will first describe a multi-phase genetic programming approach in coding eye-body-coordinated motion strategies for complex environments. Then we will present an implemented *sumo* learning system that empirically validates the effectiveness of our approach in dealing with a number of difficult sumo situations.

10.1 Issues

Researchers have demonstrated various characteristics and capabilities of mobile robotic agents. For instance, considering the task of obstacle avoidance, Miglino et al. [MLN96] developed and experimentally tested an artificial neural network-based controller for a miniature mobile robot called Khepera. The controller is evolved based on a genetic algorithm. Nolfi and Parisi [NP95] have shown that the above-mentioned approach can be readily applied to develop a capability of object-picking by the robot with one arm while avoiding obstacles.

The application of genetic programming focuses on the synthesis of robot motion/control programs. Koza [Koz92] has shown the application of genetic programming in synthesizing a subsumption controller for robots capable of performing wall-following and box-moving tasks. Reynolds [Rey92, Rey94] has applied genetic programming to create control programs that enable a mobile vehicle to avoid collisions. In this chapter, we are interested in evolving coordinated behaviors, i.e., effective maneuvers, for a sumo robot with respect to certain performance requirements. This problem is in essence a problem of genetic programming that is aimed at synthesizing local motion strategies for the robot.

10.2 Dual-Agent Learning

The problem addressed in this chapter can be stated as follows: A sumo contest is to be played in a closed environment of size $42cm \times 65cm$. One player of diameter $12cm$ is required to learn necessary sumo maneuvers that would force its opponent out of the contest area. While learning, the player is allowed to communicate with an assistant responsible for passing advice to the player in order to improve its performance. Suppose that the opponent to this player is capable of showing different complex postures as well as different resistances (as simulated by changing weights) in an attempt to minimize the strength of the player. Our goal is to develop a working sumo learning system involving a physically embodied robotic agent as the player and a computational agent as the assistant. Figure 10.1 shows a sumo contest scenario involving two mobile robots.

FIGURE 10.1. A scenario of sumo contest between two robotic agents (©1999 IEEE).

10.3 Specialized Roles in a Dual-Agent System

The sumo learning system will consist of two essential agents: the sumo player (named `Junior`) and the assistant (named `Pal`). `Junior` is a physical micro-mobile robot, as shown in Figure 10.2, whose objective is to learn sumo tricks while actually performing strategic motion maneuvers against its opponent and following the advice of `Pal`. During the training sessions, the opponent is replaced by dummy players of various postures. `Pal` is a computer system that constantly monitors and evaluates the performance of `Junior` and thereafter offers friendly advice to `Junior`. The advice that `Pal` comes up with should not overwhelm `Junior`. In other words, `Junior` should learn the general maneuvers first and then move on to the specialized ones for the sake of effective learning.

10.4 The Basic Capabilities of the Robot Agent

In the sumo learning, `Junior` can readily utilize its two arms, two eyes, a real-time action controller, memory, and a communication device. This will offer `Junior` a number of basic capabilities. For instance, the arms can effectively push and at the same time sense its opponent. They are physically implemented with two micro-switches. The two infrared sensor-based eyes are capable of detecting the presence of the opponent. In addition to the two front eyes, `Junior` is also equipped with one pair of downward-looking infrared sensors to detect whether it is off the boundary of the contest area. The real-time action controller will enable `Junior` to perform a specific maneuver with its two arms and two wheels

FIGURE 10.2. The appearance of Junior. It is equipped with two arms (Circle 1), two eyes (Circle 2), a real-time motion controller, memory, and a communication device (Circle 3) (©1999 IEEE).

(located right below its eyes). The memory will help record the present sensory information and recall a certain earlier acquired strategic maneuver. Last but not least, the communication infrastructure will serve as a channel of information and debriefing between Junior and Pal. That is, what Junior sees and feels will be communicated through such a channel back to Pal for its monitoring and maneuver programming. In a similar way, the maneuver selected by Pal can also be sent to Junior. The action controller, memory, and communication device are physically embodied in an onboard micro-controller remotely connected to Pal.

10.5 The Rationale of the Advice-Giving Agent

Pal is responsible for providing just-in-time advice to Junior. It works with a set of basic actions that Junior has and evaluates and corrects Junior's maneuvers in order to achieve a better performance. In doing so, it utilizes a multi-phase genetic programming approach that attempts to efficiently achieve the learning of general maneuvers prior to the specialized ones.

10.5.1 The Basic Actions: Learning Prerequisites

The learning prerequisites for `Junior` consist of the following actions: forward move, backward move, left turn, right turn, and stop. These actions are the basic motions that `Junior` must be able to initially perform by controlling the angular displacements of its two wheels as well as their directions.

10.5.2 Genetic Programming of General Maneuvers

In order to give just-in-time advice on general sumo maneuvers, `Pal` selects a series of basic actions in response to the sensory data as obtained and sent by `Junior` at each time step t. A sequence of such selected basic actions is viewed as a single *maneuver*. The action selection process is governed by the principle of genetic programming.

Specifically, `Pal` represents, in the form of a chromosome, a sequence of l basic actions as a candidate maneuver with respect to a certain category of sensory stimulus as received by `Junior`. In order to evolve the most effective maneuver, i.e., the advice for `Junior`, it maintains a population of such candidate maneuvers and applies the genetic operations of crossover and mutation among them. Next, it passes the candidate action sequences to `Junior` for execution and evaluates the effectiveness of resulting performance in `Junior` by calculating a fitness function. The highly effective action sequences are kept in the population for further selection. Here, the fitness function corresponds to a criterion for good sumo performance, which is specifically defined as follows:

$$S = \sum_{i=1}^{l} w_i \cdot \left(\sum_{j=1}^{n} \Upsilon^j(B_i) + \sum_{k=1}^{m} (\vartheta^k(B_i) + \varrho^k(B_i)) \right), \tag{10.1}$$

where $\Upsilon^j(B_i)$, $\vartheta^k(B_i)$, and $\varrho^k(B_i)$ will return one if `Junior` can see its opponent with side j sensing, arm k holding onto its opponent, and arm k coming in contact with the opponent, respectively, during basic action B_i. Such information is readily obtained by `Junior` via its sensors. l, n, and m denote the number of basic actions selected (which is set to 4 in our experiments), the number of eyes, and the number of arms, respectively. w_i denotes a weight for basic action B_i.

In other words, here a good performance by `Junior` entails the one in which it can constantly face and see its opponent with both-side sensing. At the same time, its two arms should firmly hold onto the body of its opponent, while pushing, until the opponent is forced out of the sumo area. Furthermore, in order to achieve a greater impact, `Junior` should try to actively engage its opponent with its arms.

10.5.3 Genetic Programming of Specialized Strategic Maneuvers

Once a certain general sumo maneuver is acquired by `Junior`, `Pal` embarks on a new course of genetic programming for `Junior` that aims to further improve the performance of `Junior` by fine-tuning the basic actions. The fine-tuning actions are created drawing on the basic ones. Table 10.1 provides a taxonomy of such actions as used in this phase of genetic programming.

Basic actions for creating general sumo maneuvers	**Fine-tuning actions for composing specialized strategic maneuvers**
`F`: forward move	`FF`: fast forward move
	`MF`: moderate forward move
	`SF`: slow forward move
`B`: backward move	`FB`: fast backward move
	`MB`: moderate backward move
	`SB`: slow backward move
`L`: left turn	`LSRF`: r-wheel forward move with stall l-wheel
	`LBRS`: l-wheel backward move with stall r-wheel
	`LBRF`: l-wheel backward and r-wheel forward move
`R`: right turn	`LFRS`: l-wheel forward move with stall r-wheel
	`LSRB`: r-wheel backward move with stall l-wheel
	`LFRB`: l-wheel forward and r-wheel backward move
`S`: stop	`S`: stop

TABLE 10.1. The basic and fine-tuning actions of which the general sumo maneuvers and the specialized strategic maneuvers are composed, respectively.

In the genetic programming of specialized strategic maneuvers, `Pal` classifies the situations that `Junior` is facing by considering not only the present sensory state but also the previous one. Accordingly, it represents candidate specialized strategic maneuvers in the form of further fine-grained chromosomes based on those of general sumo maneuvers. Such a representation incorporates the information on what sequence of fine-tuned actions should be involved. The actual implementation of genetic programming for this phase is similar to the one for the general maneuvers.

Figure 10.3 presents a schematic diagram of the multi-phase genetic programming approach that attempts to efficiently achieve the learning of sophisticated maneuvers by `Junior`.

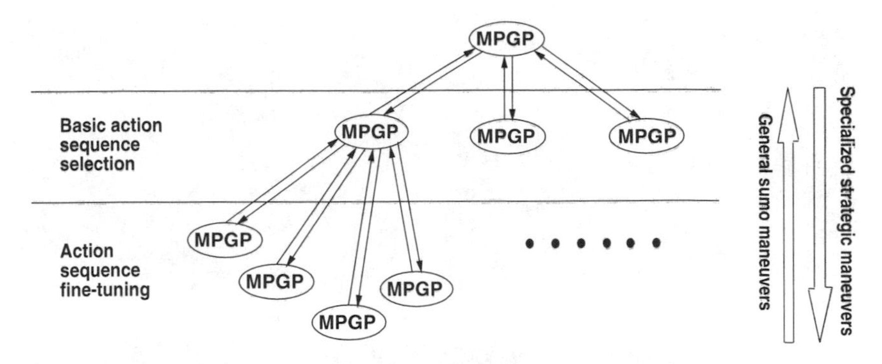

FIGURE 10.3. An illustration of multi-phase genetic programming (MPGP) as used by `Pal` in generating its advice (©1999 IEEE).

10.6 Acquiring Complex Maneuvers

Having presented the basic principles behind the dual-agent system for learning sumo maneuvers, the following sections describe the experimental verification of this approach.

10.6.1 Experimental Design

In order to demonstrate the effectiveness of learning, we have decided not to use a physical mobile robot as an opponent to `Junior`, but instead to place a number of dummy players in different sizes and shapes, corresponding to different weights and postures that a real opponent may hold against `Junior`. Figures 10.4(a)-(d) show four such postures, named flat, curved, corner, and circular postures, respectively. The aim of these experiments is to show whether `Junior`, with the coaching assistance of `Pal`, can successfully acquire eye-body-coordinated maneuvers in a complex sumo environment with opponents of varying weights and postures.

10.6.2 The Complexity of Robot Environments

There are five dummy opponents used in the experiments. Four postures are of particular interest as shown in Figure 10.4. The complexity of the dummy opponents that `Junior` faces can be viewed with respect to the weights and the *characteristic* engagement types. A classification of this complexity is shown in Figure 10.5. The vertical axis corresponds to the weight of an opponent, whereas the horizontal axis corresponds to three types of engagement contact, namely, surface, corner, and point contact. The weight difference of an opponent is considered only in the case of a flat posture.

In a surface contact, it is relatively easier for `Junior` to see its opponent with both-side sensing (i.e., with both front eyes) and at the same time engage the opponent with both arms. On the other hand, in the case of a corner contact, it is

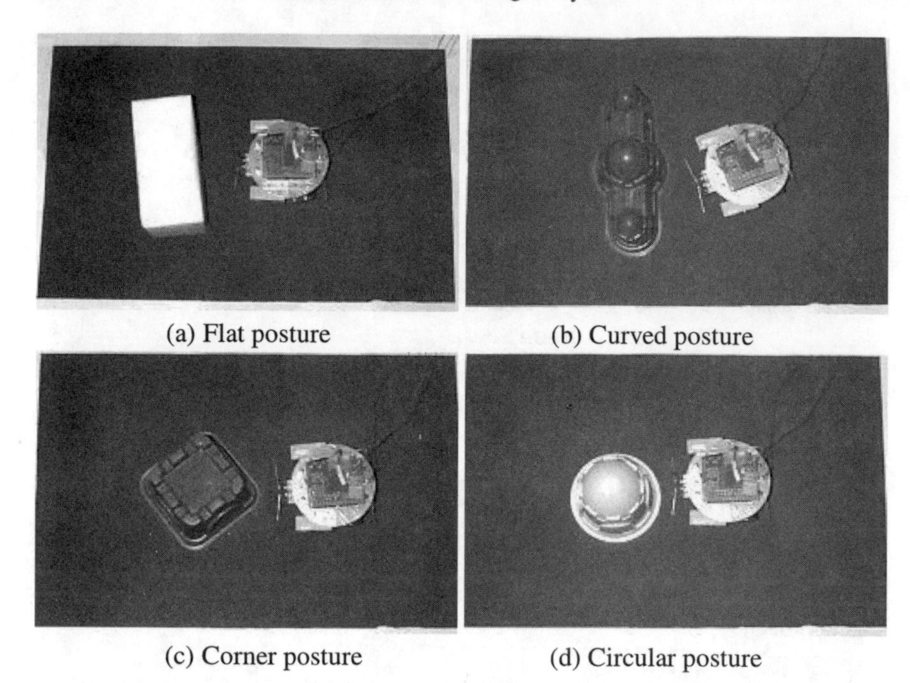

(a) Flat posture (b) Curved posture

(c) Corner posture (d) Circular posture

FIGURE 10.4. The dummy opponents in a variety of difficult postures (©1999 IEEE).

difficult for `Junior` to focus with both-side sensing and to engage with both arms due to the small size of its opponent. In addition, what makes the situation more complex is that, once `Junior` pushes its opponent with one arm, the opponent may move in an uncertain direction. When a corner contact and push occurs, this leads to an edge contact. Finally, the point contact is similar to the corner contact, except that it is impossible to have both-side sensing and both-arm engagement.

10.6.3 Experimental Results

The performance of the dual-agent sumo learning system is shown in Figure 10.6, where five specialized strategic maneuvers are genetically selected with respect to five different categories of opponent situations. They are (a) lightweight flat posture, (b) heavyweight flat posture, (c) lightweight curved posture, (d) lightweight corner posture, and (e) lightweight circular posture. The resulting maneuvers are, respectively, called (a) straight push, (b) intermittent impact, (c) stall turning, (d) corner-to-edge alignment, and (e) alternating twist. Details on the multi-phase genetic programming of these maneuvers can be viewed from the corresponding fitness-function curves as given in Figure 10.7. In Figure 10.7, dotted circle, dashed star, and solid lines correspond to the learning of coordinated maneuvers in response to left-side sensing, right-side sensing, and both-side sensing of the opponent, respectively.

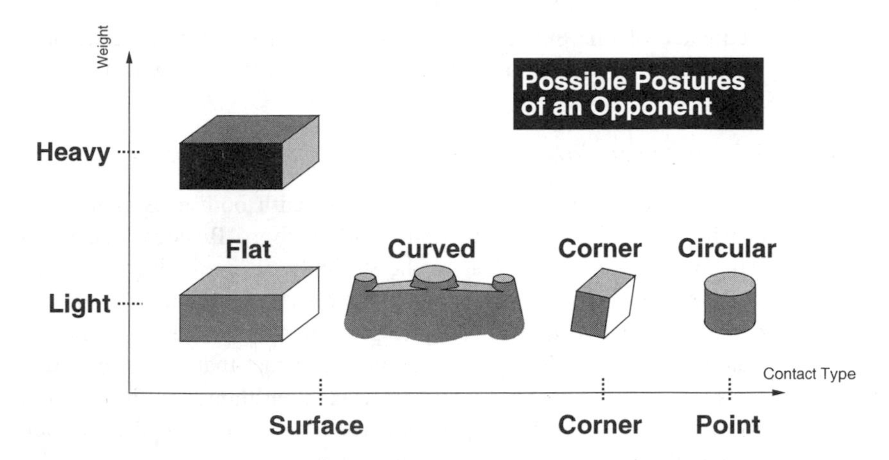

FIGURE 10.5. The complexity of the `Junior` environments, as classified according to the weights and the possible engagement contact types of an opponent (©1999 IEEE).

10.6.4 *Lightweight or Heavyweight Flat Posture*

As shown in Figures 10.6 and 10.7(a)-(b), while dealing with the lightweight or heavyweight flat posture opponent, `Junior` can easily see its opponent with both-side sensing. The fact that the dotted and dashed fitness-function curves are much shorter than the solid curve indicates that `Junior` can quickly figure out the best way to turn to its opponent with both-side sensing and thereafter concentrate on the learning of effective specialized strategic maneuvers of straight push or intermittent impact. Also it can be noted that such learning can be readily achieved once the single-side sensing is corrected after 13 genetic programming steps.

Both lightweight and heavyweight cases are similar in terms of switching from the single-side sensing mode to the both-side in learning specialized maneuvers. The key difference between them is that the maneuvers for the single-side sensing in the latter case are not as effective as those in the former case, judging from their fitness-function curves. In addition, the strategic maneuvers in the latter case create a sequence of fine-tuned actions for high-impact intermittent push. In other words, the multi-phase genetic programming by `Pal` has found a niche maneuver of intermittent impact that could score high fitness values in the situation of a heavyweight opponent.

10.6.5 *Lightweight Curved Posture*

Concerning the case of lightweight curved posture opponent, the resulting maneuver transitions from a single-side sensing mode to a both-side sensing mode, *repeatedly*. This is reflected in the concurrent learning of three sensing mode maneuvers as shown in Figure 10.7(c). The solid line in the figure indicates that the selected actions are similar to those in the lightweight flat posture case. As a result, a sequence of fine-tuned actions is selected that creates a behavior of effective stall turning, as an adaptation to the curved posture opponent.

10.6.6 Lightweight Corner Posture

In this case, the both-side sensing of an opponent with both eyes is impossible. This is why the solid fitness-function curve is short. Besides, due to the uncertainty in the environment – that is, the difficulty in predicting the movement directions of the opponent – the maneuvers for the situation of single-side sensing are not as effective as those in the preceding three cases, as may be compared based on their fitness-function curves. The fact that the fitness function in this case can be stabilized at a certain level, even though without further improvement, reveals that the basic performance of `Junior` is still achievable using the genetically programmed specialized maneuvers.

10.6.7 Lightweight Point Posture

As shown in Figure 10.7(e), the maneuvers acquired are slightly more effective than those in the previous case, meaning that the movement uncertainty in the point posture is less than that in the corner posture, since it does not involve any engagement contact transitions. As a result, the multi-phase genetic programming has led to a niche maneuver of alternating two arms.

10.7 Summary

This chapter has described a coordinated maneuver learning system in which one agent named `Junior` senses its environment and performs physical maneuvers. Another agent named `Pal` offers just-in-time suggestions for improving performance by way of monitoring and generating better and specialized candidate maneuvers. The two agents work collaboratively by offering either performance feedback or maneuver advice. In order to test the adaptability of sumo behaviors, we created various complex environments for `Junior` by presenting a number of dummy opponents of varying weights and postures.

From our experiments, it is demonstrated that the coaching assistance provided by `Pal` can gracefully lead `Junior` to become a competent player who executes specialized maneuvers. The key to `Pal`'s coaching lies in the use of a real-time multi-phase genetic programming approach. The developed genetic programming approach is effective in searching for niche solutions in terms of both general and specialized maneuvers for `Junior` to deal with a variety of situations, even in the presence of uncertainty in the dynamics of its opponent. Furthermore, from the obtained results, it is shown that the multi-phase genetic programming approach is an efficient way of learning by focusing on general solutions (e.g., general sumo maneuvers) first and then moving on to specialized ones (e.g., specialized strategic maneuvers) based on the obtained high-fitness general solutions (e.g., inheriting general sumo expertise).

While comparing the multi-phase genetic programming approach to the single-phase one, it can be noted that the former does not involve *all* sensory and control variables. Instead it groups similar ones together in order to speed up learning. Once the learning in this phase is achieved, it divides the groups to consider more variables in a given situation.

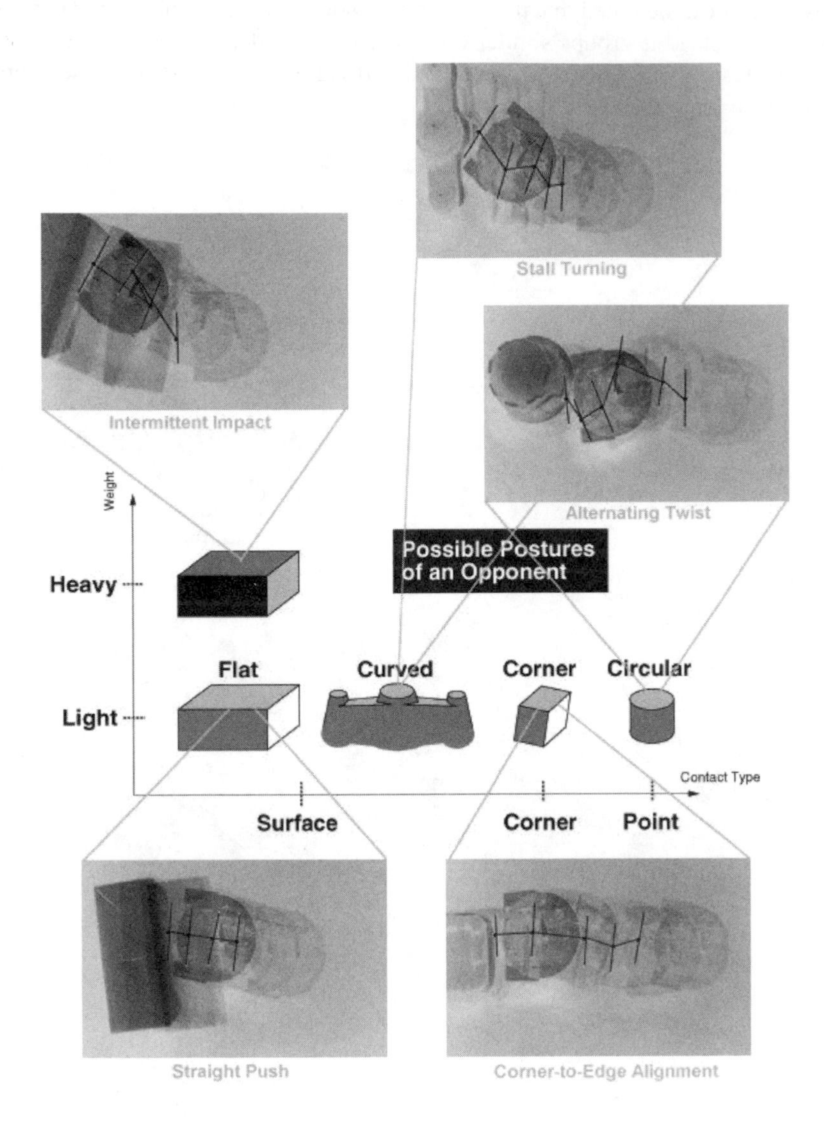

FIGURE 10.6. The specialized strategic maneuvers resulting from multi-phase genetic programming (©1999 IEEE).

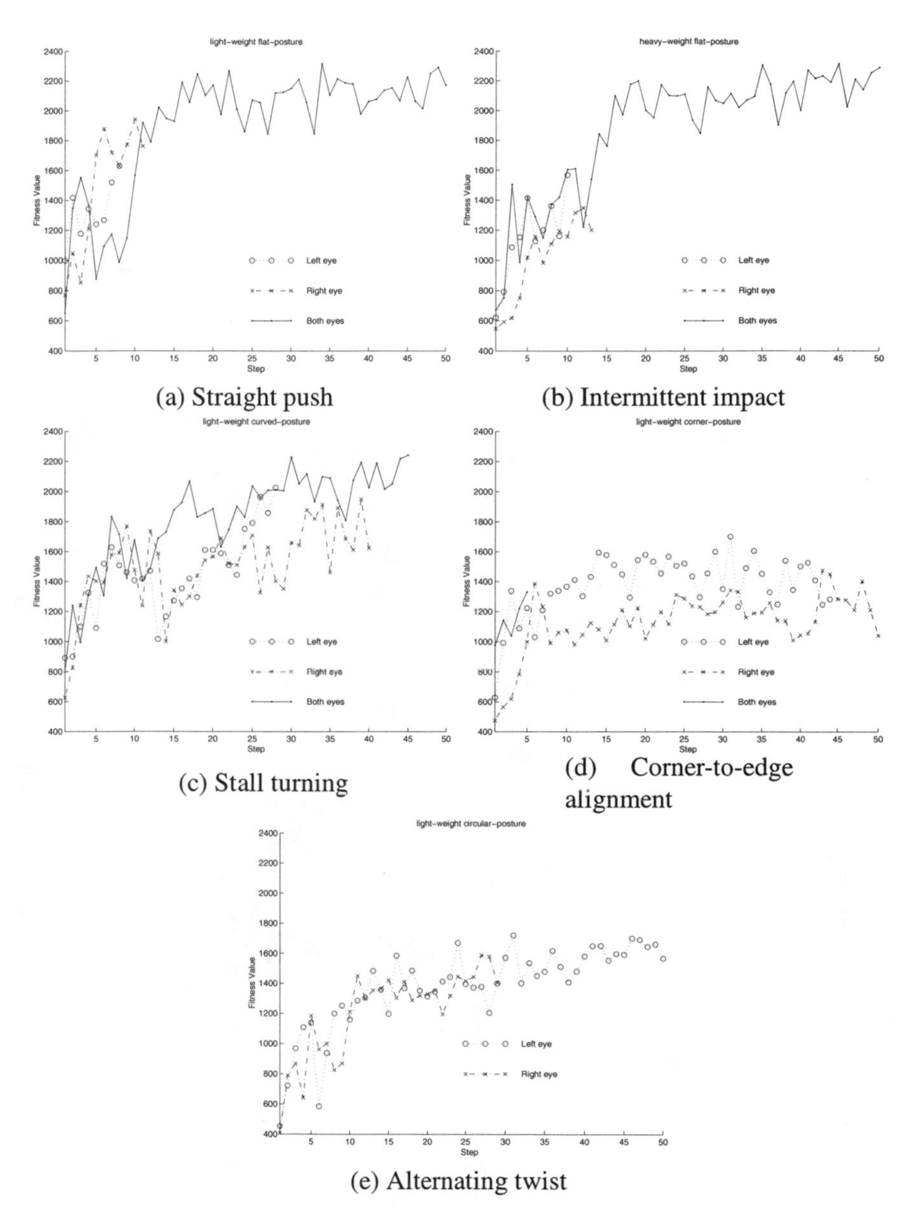

FIGURE 10.7. Fitness-function curves recorded during the multi-phase genetic programming of five specialized maneuvers for `Junior` (©1999 IEEE).

11

Collective Behavior

If the observer were intelligent (and extraterrestrial observers are always presumed to be intelligent) he would conclude that the earth is inhabited by a few very large organisms whose individual parts are subordinate to a central directing force. He might not be able to find any central brain or other controlling unit, but human biologists have the same difficulty when they try to analyze an ant hill. The individual ants are not impressive objects – in fact they are rather stupid, even for insects – but the colony as a whole behaves with striking intelligence.[1]

Jonathan Norton Leonard

Decentralized autonomous robots can effectively perform certain tasks with their collective behaviors. In this chapter, we will address the problem of how to

[1] Other-Worldly Life, *Flight into Space*, Sidgwick & Jackson Ltd. and Random House Inc., 1953.

acquire a collective goal-directed, task-driven behavior in a group of distributed robots. The specific task that we will consider here is for the individual robots to perform coordinated movements that will push a box toward a desired goal location. The difficulty of this task is that the movements of the robots with respect to the box cannot follow an explicit, global plan and control command, due to certain modeling limitations as well as planning costs. In such a case, it is important that the individual robots locally modify their movement strategies. At the same time they must create a desired collective interaction between the robot group and the box, which will successfully bring the box to the goal location.

In order to solve this problem, we will present an evolutionary computation approach in which no centralized modeling and control are involved except a high-level criterion for measuring the quality of collective task performance. The evolutionary learning approach is based on a fittest-preserved genetic algorithm (GA). Our approach begins with a model of local interactions between the robots and the box in the environment. We then apply a GA as a global optimization technique to select the local motion strategies of the individual robots in an attempt to maximize the overall effectiveness of moving the box toward the goal location.

While giving the formulation and algorithm of the GA-based collective behavior learning, we will also provide the results of several computer simulations for illustrating and validating the effectiveness of our approach. In order to further examine the convergence properties of the group behavior learning, we will use a Markov chain to model the genetic algorithm and to derive the probability of success for the distributed robots to obtain their collective goal-directed behavior after some iterations of evolutionary computation.

The chapter is structured as follows: Section 11.1 provides a formal statement of the problem addressed. Section 11.2 gives an overview of the approach to group behavior learning with both modeling and algorithmic details. Sections 11.3 and 11.4 describe two sets of computer simulations. The first illustrates how the group behavior learning works, with a simplified dynamics model of robot-box interaction. The second presents some physically modeled, collective box-pushing examples to further investigate the effectiveness of learning in a more complex pushing task, where robots of rectangular shape can have many possible point contacts while pushing and orienting a box. Section 11.5 describes a Markov chain model of the GA-based behavior learning and examines its convergence properties. Section 11.6 summarizes this chapter.

11.1 Group Behavior

Many existing approaches to robot motion planning have, to some extent, shared one thing in common; namely, the motion of robots is planned and executed by utilizing complete and/or partial representations of a task environment, such as box shape, obstacle dimensions, and the position and orientation information about other robots. The question that remains is how to effectively generate coordinated motions for a group of robots when the information about the environment, such as the states of other robots, is not available or too costly to obtain. This issue is particularly relevant if we are to develop group robots that can work collectively and adaptively on a common task, even though each robot can only sense and hence react to its environment locally.

11.1.1 What is Group Behavior?

Let ε^t denote the local environment state of a robot group at time step t and ξ denote the sensory inputs to the individual robots. In the environment, each robot is capable of performing some actions τ. A coordinated movement, U, by the group robots in response to ε^t is composed of actions τ such that they satisfy μ, the constraints imposed by the environment such as point contacts with a box, and ν, the criterion of collective task performance. This may be expressed as follows:

$$\forall \varepsilon^t,\ \exists \gamma^t \in \tau,\ \ s.t.\ U \equiv \{\ \gamma^t\ \},\ \ E(\mathcal{A}) \stackrel{U}{\Longrightarrow} \nu, \tag{11.1}$$

where $\{\gamma^t\}$ corresponds to a goal-directed group behavior at time step t. $\mathcal{A}$ denotes group robots. $E(\mathcal{A})$ denotes the performance evaluation of $\mathcal{A}$.

11.1.2 Group Behavior Learning Revisited

The problem of group behavior learning is essentially a problem of learning how to collectively perform a given task. Solving this problem requires group robots to continuously gain experience from their local interaction with an environment. Such group experience is incrementally updated. This may be expressed as follows:

$$\epsilon^t\ =\ \varepsilon^{t-1} \oplus \epsilon^{t-1},\qquad U \text{ leads to } \epsilon^t\ \mid_{\Delta\epsilon^t \approx 0,\ E(\mathcal{A})=\nu}, \tag{11.2}$$

where ϵ^t represents the group experience at time step t, $\oplus$ denotes a computational updating mechanism, and ε^{t-1} corresponds to the environment state sensed at time step $t-1$.

A group behavior will be generated from group experience following a series of environment state transitions. Therefore, the process of group behavior learning may be regarded as the process of converging group experience.

The specific group learning problem to be considered in this chapter can be stated as follows: given a group of three individual robots, $\mathcal{A}$, capable of sensing and changing their own local positions and orientations, ε^t, with respect to a box and a desired goal location in their workspace, how to develop a genetic algorithm-based mechanism, $\oplus$, such that $\mathcal{A}$ can gradually acquire a coordinated movement, U, based on a series of ϵ^t changes. It is required that as a result of U, the box will be collectively pushed by $\mathcal{A}$ toward a goal location, $\varepsilon^{t'}$.

As in some real-world situations, here we assume that each group robot has limited sensing, communication, and onboard computation capabilities. More specifically:

1. **Sensing:** The group robots are capable of distinguishing which is the box to be pushed and which is its group mate (e.g., through color detection). Due to the limitation in its sensing range, the robot will not be able to detect the exact shape of the box. The robot can only sense the direction of the common goal location – hence estimate its relative orientation – but not the exact distance from the goal.

2. **Communication:** There will be no cross-communication between group robots. Each group robot can only establish a communication channel with a remote evolutionary-computation agent for receiving the next local movement command.

3. **Onboard computation:** All coordinated movement evaluation and selection are handled by a remote agent. The onboard computation of a robot is responsible only for its motor-level control, local sensing, and communication.

Besides the above assumptions, to focus on the issue of learning collective box-pushing behavior, we assume that each group robot is capable of finding the box, in case the contact between the robot and the box is lost. We will demonstrate our approach with a series of simulations in which we ignore modeling and control uncertainties. The observations and experience gained from such simulation-based studies can provide us with insights into group behavior learning in physical robotic systems.

11.2 The Approach

This section presents a genetic algorithm-based approach to the problem of learning how to perform a collective box-pushing task by a group of distributed robots.

11.2.1 The Basic Ideas

Figure 11.1 presents a schematic diagram of the computational architecture for genetic algorithm-based group behavior learning. As shown in the figure, the fitness function of our genetic algorithm involves two components: one corresponds to the interactions (the relative spatial configurations of individual robots relative to a box), whereas the other corresponds to global feedback information (the direction of a net pushing force on the box relative to a goal direction, as a result of robot spatial configurations). Based on the current spatial configurations of the robots, the genetic algorithm will first generate a set of new spatial configurations,

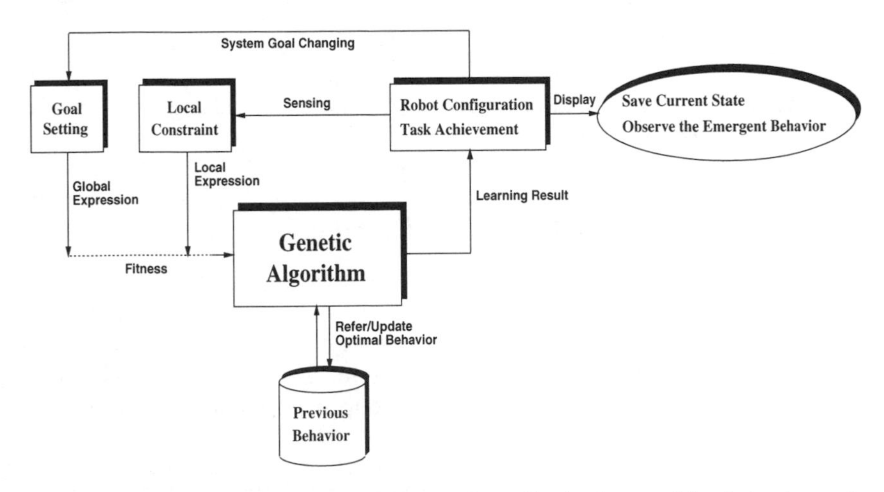

FIGURE 11.1. The computational architecture for collective box-pushing behavior evolution. The movements of group robots in terms of their local spatial relationships, such as a relative orientation with respect to a box being pushed and a desired goal location, are gradually selected by a genetic algorithm-based evolutionary computation mechanism that computes and evaluates the outcomes of potential motion strategies for the robots (©1999 IEEE).

evaluate these configurations using the fitness function (as denoted by a dashed line), and then select the high-fitness configurations for the next step of evolution. In doing so, the local motion strategies for the robots are continuously evolved and performed. Each strategy is encoded in terms of the relative positions and orientations of the robots.

11.2.2 Group Robots

In our work, simulated mobile robots will be used, which are graphically represented as circular or rectangular shapes[2]. Each robot is equipped with an array of ultrasonic sensors to sense a box of a certain height in its workspace environment.

11.2.3 Performance Criterion for Collective Box-Pushing

The task performance of group robots in collective box-pushing will be evaluated based on a high level criterion. This criterion favors the following collective performance:

1. The robots maximize their net pushing force on the box.

[2]Here we consider rectangle-shaped robots in order to make the box-pushing problem more complex and interesting – in the rectangular case, the configurations of the contacts between a robot and a box being pushed will not be unique.

2. The direction of the box movement is pointed toward a desired goal.
3. Whenever a goal location is redefined, the robots change their motion strategies accordingly.

11.2.4 Evolving a Collective Box-Pushing Behavior

The mechanism for selecting robot movement strategies is based on a fittest-preserved genetic algorithm (GA). This algorithm emphasizes on the highest-fitness individual[3], i.e., the best movement strategy, by recording it to the next generation and applying mutation to this individual without any crossover operation. The complete GA as used in the group behavior learning is given in Figure 11.2.

begin
  define fitness function S_f,
  define the maximum number of generations per step $\mathcal{G}$,
  define population size $\mathcal{P}$,
  define crossover probability p_c,
  define mutation probability p_m,
  create an initial population of $\mathcal{P}$ members,
  for $generation : 1 \longrightarrow \mathcal{G}$ **do**
    evaluate fitness S_f in the current $generation$:
      for $population : 1 \longrightarrow \mathcal{P}$ **do**
        propose new spatial configurations for group robots,
        compute the corresponding motions of the group robots,
        evaluate the fitness values for the computed movements,
      endfor
    record the highest-fitness individual,
    mutate the highest-fitness individual with p_m to get a
      new individual for the next *generation*,
    select additional $\mathcal{P} - 1$ individuals based on their fitness values,
    apply a two-point crossover operation to the $\mathcal{P} - 1$ individuals
      to get $\mathcal{P} - 1$ new members,
    mutate the $\mathcal{P} - 1$ new members with probability p_m
      to create the next $generation$,
  endfor
end

FIGURE 11.2. The genetic algorithm that is used to compute, evaluate, and select collective goal-attaining, box-pushing motion strategies for group robots.

[3]A distinction should be made between the individuals in a GA population and the individual robots in a group.

11.2.5 The Remote Evolutionary Computation Agent

The entire system for a collective box-pushing task is composed of two parts; namely, distributed group robots that directly interact with a task environment and a remote agent that handles the computational and communication duties required for the task. The remote agent will be responsible for performing the GA-based evolutionary computation and broadcasting selected movement strategies to the group robots.

11.3 Collective Box-Pushing by Applying Repulsive Forces

In order to illustrate the key steps of the above-mentioned group behavior learning approach and to show its general features, this section presents the formulation and experimentation in an illustrative collective box-pushing example. In this example, we assume that the robots can be regarded as point robots. In addition, we define the dynamics of robot-box interaction by introducing a notion of artificial repulsive force.

11.3.1 A Model of Artificial Repulsive Forces

The model of artificial repulsive forces between group robots and a box is based on a Hooke's law, spring-like model as follows: if a robot moves inside an l_0-radius region around the box, it will immediately exert an artificial repulsive force $\vec{F_i}$ on the box. The direction of $\vec{F_i}$ is shown in Figure 11.3, and the scale of $\vec{F_i}$ is computed as follows:

$$F_i = \begin{cases} \eta \cdot (l_0 - d_i), & \text{if } d_i \leq l_0, \\ 0, & \text{otherwise}, \end{cases} \tag{11.3}$$

where η is a positive coefficient, and d_i is the Euclidean distance between the robot and the box at the time of exerting a repulsive pushing force. If the robot is outside the l_0 range of the box, the pushing force will become zero.

By introducing a notion of artificial repulsive forces, we will regard the interaction between three group robots and a box as being created by such forces. In other words, the group robots will collectively push the box through their repulsive forces, if and only if their distances to the box are less than or equal to a predefined constant, l_0.

11.3.2 Pushing Force and the Resulting Motion of a Box

In this example, we are interested in the coordination among three group robots in pushing a box toward a desired goal location. The pushing force acting on

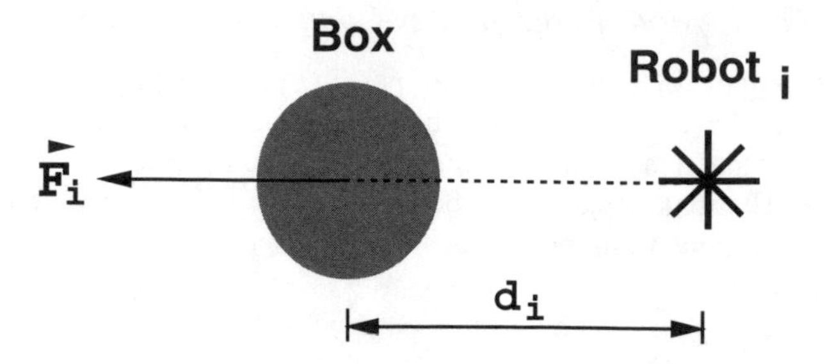

FIGURE 11.3. A top view of collective box-pushing (Note: for the sake of clarity, only one robot is depicted) where group robots, marked with *, move toward a box, signified by a shaded area. When their distances d_i are less than or equal to threshold l_0, pushing forces will be exerted on the box by the robots, according to the model of a repulsive pushing force (refer to the text).

the box will be contributed by all three robots. Here, we apply the rule of vector composition to combine the pushing forces as created by the robots.

We consider that the motion of a box is caused by the collective net repulsive pushing force on the box, $\texttt{vector\_sum}(\vec{F_i})$, as follows: the direction of box displacement is the same as that of the net force, and the magnitude of box displacement is proportional to that of the net force.

11.3.3 Chromosome Representation

In the genetic algorithm as given in Figure 11.2, each of the members in a population is encoded using a gray code, called a chromosome. Specifically, we assign an 8-bit substring to encode the relative spatial configuration of one of the three robots, as shown in Figure 11.4(a). Furthermore, we subdivide the 8-bit substring into two parts, as in Figure 11.4(b), that correspond to orientation angle β_i and distance d_i of a robot relative to a box, respectively, as illustrated in Figure 11.5. The most significant 5 bits of the 8-bit substring are assigned to encode β_i. That is, the entire range of 2π for the relative orientation is divided into 32 *units* with an angular resolution of $\frac{\pi}{16}$. The other 3 bits are assigned to encode d_i. Thus, the distance is divided into 8 *units* with a linear resolution of $\frac{l_0}{8}$.

It may be noted from Figure 11.5 that β_i and d_i together form a polar coordinate for the robot. In this polar coordinate system, the polar axis is originated at a certain reference point on the box, pointing toward the goal, as denoted by +. Thus, if robot i has distance d_i from the box and orientation β_i in this polar coordinate system, then the relative spatial configuration of robot i can be expressed as follows:

$$\vec{T_i} = d_i \cdot e^{j\beta_i}. \tag{11.4}$$

Accordingly, the direction of a pushing force, $\vec{F_i}$, as produced by the robot will become $e^{-j\beta_i}$. The magnitude of this force will satisfy Eq. 11.3. Thus, we have:

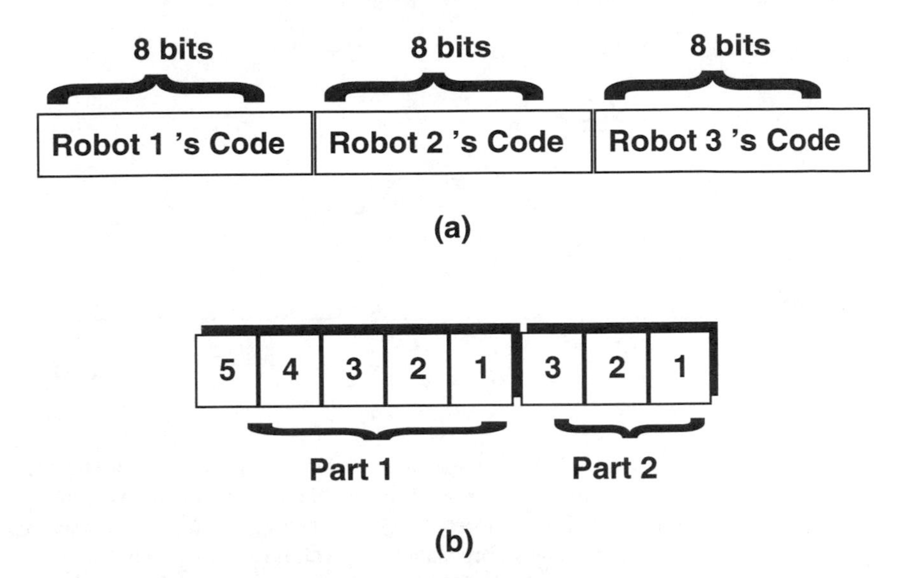

FIGURE 11.4. (a) A 24-bit binary string is used to represent individual members, i.e., chromosomes, in the genetic algorithm-based motion strategy selection for three group robots. (b) Each of the 8-bit substring in the chromosome, encoding the local motion strategy of a single robot, consists of two parts: part 1 encodes relative orientation angle β_i and part 2 local distance d_i (©1999 IEEE).

$$\vec{F_i} = F_i \cdot e^{-j\beta_i}. \tag{11.5}$$

11.3.4 Fitness Function

In the genetic algorithm as used for evolving robot coordinated movement strategies, a function is used to continuously evaluate the fitness of each member (a potential strategy) in a population, which is referred to as a fitness function. This function represents a high-level criterion, as stated earlier, for the task of collective box-pushing. Specifically, our fitness function is defined as follows:

$$S_f = s_1 \cdot s_2 \cdot s_3, \tag{11.6}$$

where s_1 specifies a net pushing force on a box as created by group robots, which can be mathematically expressed as follows:

$$s_1 = \alpha \cdot ||\texttt{vector\_sum}_i(\vec{F_i})||. \tag{11.7}$$

s_2 measures how close the robots are to the box, which is calculated as follows:

$$s_2 = \kappa \cdot [\sum_i d_i]^{-1}. \tag{11.8}$$

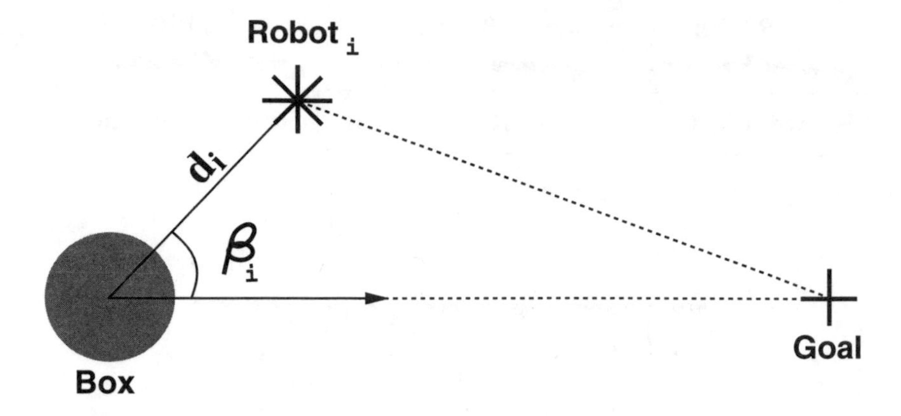

FIGURE 11.5. A top view of robot local spatial configuration in a movement strategy (Note: for the sake of clarity, only one robot, marked with *, is shown). In our experiment, once a robot has moved to a location close enough to a box ($d_i \leq l_0$), it can exert a repulsive force on the box at its relative orientation, β_i, with respect to a reference point on the box and a desired goal location. β_i and d_i can be regarded as the polar coordinates of the robot in a coordinate system defined by the current and the desired locations of the box (©1999 IEEE).

Finally, s_3 indicates whether and how much the motion of the box is directed toward a desired goal location, or more specifically:

$$s_3 = \begin{cases} cos\Theta, & \Theta \in [-\frac{\pi}{2}, \frac{\pi}{2}], \\ 0, & \text{otherwise}, \end{cases} \tag{11.9}$$

where Θ denotes the direction of the net force by the three robots.

11.3.5 Examples

In this section, we will describe some experiments in which the GA-based group behavior learning mechanism is implemented and validated. In order to have a closer examination into the process of coordinated movement, we will present and trace several snapshots of the behavior selection and motion execution in a typical case study. In our experiments, we assume that the robots exert repulsive forces simultaneously.

11.3.5.1 Task Environment

In our experiments, we define the workspace for three group robots as a square arena, $\mathcal{E}$, of size $10\, l_0 \times 10\, l_0$. Whenever the robots or the box hit the boundary of the workspace, the task of box-pushing will be reset. At the beginning, the three robots and the box will be randomly placed inside $\mathcal{E}$. A desired goal location will be arbitrarily generated within the workspace.

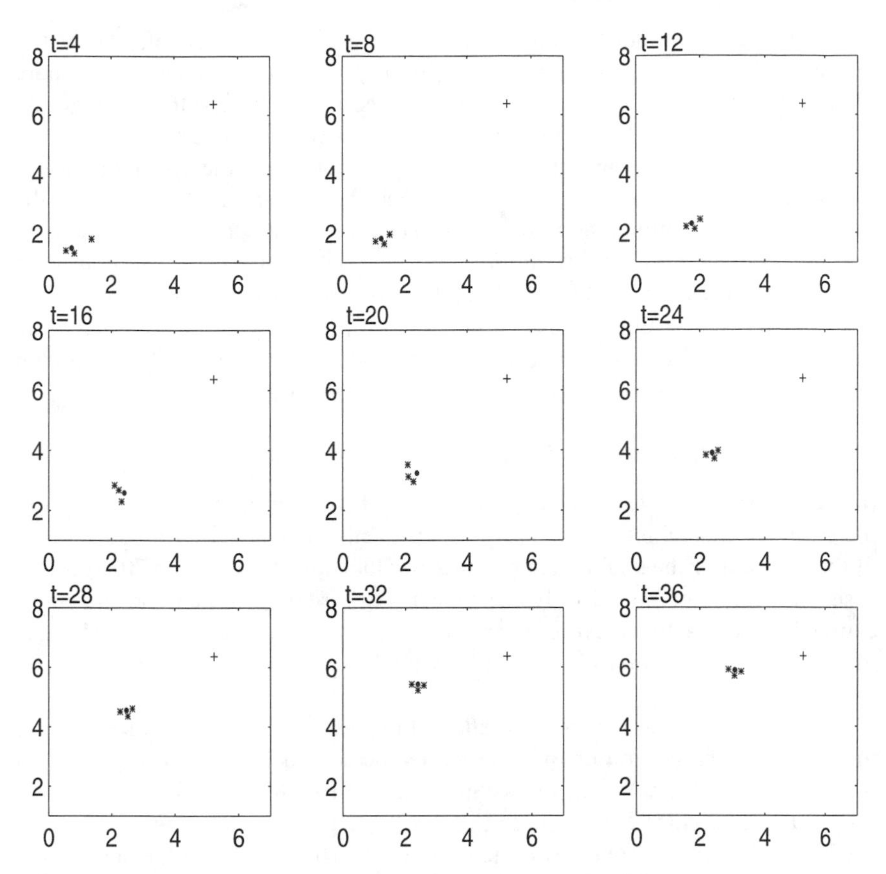

FIGURE 11.6. The snapshots of collective box-pushing by three robots taken every 4 generations from generations 4 to 36 (top view). The collective movement in each snapshot is generated based on *the fittest member of a population* at a certain generation of collective behavior evolution. In the robot environment, a desired goal location is marked with symbol +. A box is marked with symbol • (©1999 IEEE).

11.3.5.2 Simulation Results

Figures 11.6 and 11.7 provide a series of snapshots showing the process of group behavior learning based on a genetic algorithm, from generations 4 to 72 (taken every 4 generations). From these snapshots, we note that three robots continuously modify their spatial configurations in order to satisfy a high-level fitness requirement. As a result, a box, marked with •, is gradually moved toward a desired goal location, marked with +. The box reaches the goal location at generation 64.

11.3.5.3 Generation of Collective Pushing Behavior

Figure 11.8 presents the trajectories of three robots pushing a box. In the figure, the solid line denotes the trajectory of the box, and the dashed line, dotted line and dash-dot line correspond to those of the three robots. At the beginning of box-pushing, the robots move in a rather chaotic manner. It took more than 20 generations for the group of distributed robots to develop (emerge) an effective group behavior that maximizes the fitness function as given in Eq. 11.6. The motion trajectories are gradually getting smoother, as the evolution of collective motion strategies goes on. After about 40 generations, as marked in the figure, the coordination among the robots becomes most apparent. We also note that the group creates a *spiral motion* near the goal location. This is because in many high-fitness individuals selected (effective strategies), the direction of a net repulsive force may not be directly pointing toward the goal location, implying that the GA-based learning after the given period of evolution yields a good, but not necessarily the best, strategy for accomplishing the task.

11.3.5.4 Adaptation to New Goals

In our experiments, we are also interested in whether group robots can adapt to the change of a goal location. Figure 11.9 presents the traced motion trajectories of the robots after the goal is changed to a new location at generation 70. In such a case, the robots can immediately bring the box toward the new goal location, indicating that the GA-based evolutionary behavior learning can adapt to the changes in the task environment. In Figure 11.9, symbol $\circ$ at t=70 denotes a previous goal location.

The above observations are also reflected in a fitness function plot for the best individual of each generation over the entire process of behavior learning, as given in Figure 11.10. From the figure, we may identify four distinct stages in the evolution. The first stage occurs during the early 20 generations, where the robots execute their motion strategies quite chaotically. The next stage happens in the following 19 generations – from generations 21 to 39. During this period, the robots execute several collective group behaviors of low fitness. Then, during later task performance from generations 40 to 64, improved collective behaviors start to emerge, even though there are a few slight oscillations involved. Finally, after generation 64, the fitness values reach a new level, even when the goal location is changed twice – one at generation 70 and another at generation 97.

11.3.5.5 Discussions

In the formulation of our fitness function, we have emphasized keeping a good balance between the local interaction of individual robots and the global goal-directedness of box movement. This is best reflected in the three terms of Eq. 11.6. Now let us consider what would happen if either the local interaction or the global goal-directedness were removed in the fitness function.

As we may recall, s_1 and s_2, as defined in Eqs. 11.7 and 11.8, respectively, imply that all robots should try to surround the box as tightly as possible while maintaining a maximum pushing force. If only these two terms were involved in the task performance criterion, the robots would immediately acquire a series of pushing movements without taking into account any specific direction of pushing.

In such a case, the box would be pushed straight toward the boundary of the robot workspace – any turning can be against previously evolved behavior patterns and indicate low fitness.

Unlike s_1 and s_2, s_3 in Eq. 11.6 implies that the net pushing force as created by group robots should be pointed toward the direction of a desired goal location without considering the exact distance from the goal. This term of global goal-directedness information alone will by no means create any effective pushing co-ordination among the robots. In such a case, even though the box may be brought to the goal location eventually, the local motions of the robots will become very chaotic.

From the above discussions, we note that there should be a good balance between the overall goal-directedness and the local movement behavior.

By incorporating a constraint on the local movement behavior into the fitness function, we have embedded certain designer's knowledge in the fitness function. However, such high-level knowledge may not be complete as in many GA applications. This was also reflected in the experiment that resulted in a near-optimal solution. In the case studies, we note that although the collective box-pushing task is achievable, the fitness values appear to fluctuate due to the definition of the three-term fitness function. From Eqs. 11.7 and 11.8, some slight changes in the relative position and orientation of the group robots may produce quite dramatic changes in the group fitness value.

11.4 Collective Box-Pushing by Exerting External Contact Forces and Torques

The aim of the previous example is to illustrate how the group behavior learning approach works and the general characteristics of this approach. Toward this end, we simplify the dynamics of robot-box interaction by introducing a notion of repulsive forces and at the same time treat group robots and a box as points of zero physical size.

In some earlier studies conducted by Mataric [MNS95], it was shown that the location on the box that a robot is pushing can, to a certain extent, affect the performance of a pushing task. In this section, we will consider the problem of group behavior learning to perform a complex box-pushing task, where rectangular robots and a cylindrical/cubic box are involved that can have many possible point contacts among them. The goal of this example is to further examine and validate the effectiveness of learning in a more realistic simulation that is closer to a real-world situation than the preceding case study.

11.4.1 Interaction between Three Group Robots and a Box

Our next example involves three robots and one box. Two types of boxes are considered: one is cylindrical and another is cubic. The boxes are rigid bodies, whose motions are based on the influence of forces and torques externally applied by the group robots.

In order for group robots to move a box, the robots must simultaneously exert external forces and torques on the box. The individual forces and torques arise from the point contacts between the robots and the box. Here we disallow collisions among the robots and the box. We assume that the robot always produces a unit force if it is in contact with the box in a certain direction. The net contact force on the box by the three robots can produce a linear displacement of the box. A contact force may be applied to the box, at some distance from its center of mass, which may create a corresponding torque. If the net torque on the box is non-zero, the box will rotate.

11.4.2 *Case 1: Pushing a Cylindrical Box*

11.4.2.1 Pushing Position and Direction

In the case of pushing a cylindrical box, the contact positions and directions of robots can be described as follows: assume that there are $\mathcal{N}$ positions around the box at which the robots may contact and push, and each robot can produce a pushing force in one of $\mathcal{M}$ directions. All these locations and directions are evenly distributed around the box and the robot. Thus, the resolutions for describing the pushing positions and the pushing-force directions are $\frac{2\pi}{\mathcal{N}}$ and $\frac{2\pi}{\mathcal{M}}$, respectively.

11.4.2.2 Pushing Force and Torque

Figure 11.11 illustrates a contact pushing force on a cylindrical box that leads to its motion. Suppose that the direction of the pushing force is θ_i, and robot i pushes the box at a position with angular measurement β_i. In this case, the box will be pushed to move by unit force $\vec{F_i}$ and torque J_i, i.e.,

$$\vec{F_i} = \mathcal{K} \cdot e^{j\theta_i}, \tag{11.10}$$

$$J_i = \pm\mathcal{K} \cdot l_i, \tag{11.11}$$

where $\mathcal{K}$ is the magnitude of the unit force, and l_i is the distance from the center of mass to the direction of pushing force $\vec{F_i}$. When the direction of $\vec{F_i}$ is counter-clockwise, the sign of J_i is positive; otherwise, it is negative.

With the pushing force and torque, the box will be pushed to move in the direction of the force and, at the same time, rotate about its center. The linear and angular displacements are defined as follows:

$$\mathcal{O} = \alpha_1 \cdot ||\texttt{vector\_sum}_i(\vec{F_i})||, \tag{11.12}$$

$$R_\alpha = \alpha_2 \cdot \sum_i J_i, \tag{11.13}$$

where α_1 and α_2 are positive coefficients, and $||\texttt{vector\_sum}_i(\vec{F_i})||$ and $\sum_i J_i$ correspond to net force and net torque, respectively.

11.4.3 Case 2: Pushing a Cubic Box

As in the preceding case, group robots will push a box through their contact forces. However, the present case differs from the above in that a cubic box is used, whose orientation will no longer be unique.

11.4.3.1 The Coordinate System

Figure 11.12 presents a top view of group robots pushing a cubic box. For the sake of clarity, only one of the three robots is shown. During collective box-pushing, the relative spatial configuration of each robot is specified by a relative orientation, β_i, with respect to a desired goal location and a reference point on the box. The coordinate system for describing β_i is shown in Figure 11.12. In this system the origin lies in the reference point on the box, and the axis points toward the desired goal.

11.4.3.2 Pushing Force and Torque

Figure 11.13 shows a contact-pushing force on a cubic box, which causes the motion of the box. In this case, the definitions of force, torque, and displacements are the same as those in the preceding cylindrical box case.

11.4.4 Chromosome Representation

In the present case studies, the chromosome for each individual robot is encoded in 10 bits. Thus, each member of a population, expressing a possible collective motion strategy for three robots, is represented using a 30-bit chromosome. In each 10-bit substring, the most significant 5 bits express angle β_i for one of the three robots in a polar coordinate system, as mentioned above. That is to say, the total range 2π of orientation is subdivided into 32 units with an angular resolution of $\frac{\pi}{16}$. The remaining 5 bits encode the direction of a pushing force acting on the box by that robot.

11.4.5 Fitness Functions

Case 1: Pushing a Cylindrical Box

In this case, we measure the performance of a group behavior using the following fitness function:

$$S_f = s_1 \cdot s_2 \cdot s_3, \tag{11.14}$$

where

$$s_1 = ||\texttt{vector\_sum}_i(\vec{F_i})||, \tag{11.15}$$

$$s_2 = 1 + cos\Theta, \tag{11.16}$$

$$s_3 = ||\sum_i J_i||^{-1}, \tag{11.17}$$

where Θ is the direction of the net contact force by all three robots. $\vec{F_i}$ and J_i denote a contact force and a corresponding torque created by robot i, respectively.

In the above definition of the fitness function, s_1 implies that the box should be pushed by a large net contact force, s_2 implies that the box should be moved toward a desired goal location, and s_3 implies that the rotation of the box should be discouraged during collective pushing.

Case 2: Pushing a Cubic Box

In the case of pushing a cubic box, we calculate the fitness function for group robots as follows:

$$S_f = \sqrt{s_1} \cdot (s_2 \cdot s_3)^4, \tag{11.18}$$

where s_1, s_2 and s_3 are given in Eqs. 11.15, 11.16, and 11.17, respectively.

11.4.6 Examples

This section presents some experiments in which the above-mentioned GA-based group behavior learning is implemented and validated.

11.4.6.1 Task Environment

Our experimental environment for box-pushing is a square arena $\mathcal{E}$ of $100units \times 100units$. If a robot or a box hits the boundary of the environment, the experiment is reset. At the beginning, group robots and the box are randomly placed inside $\mathcal{E}$. The task of the three robots is to move the box to a desired goal location. The location of the goal is also randomly selected at the beginning. Whenever the box reaches the goal, a new goal location is selected.

11.4.6.2 Adaptation to New Goals

While observing the evolution of collective box-pushing behaviors in group robots, we will also study the relationship between group behavior learning for reaching one goal location and that for a different goal location. In order to do so, we will conduct an experiment in which the group robots are required to move the box toward a new goal location as soon as the box reaches its old goal location. We are interested in knowing whether the group robots can adapt to the change of a desired goal location by reselecting some previously evolved motion strategies.

11.4.6.3 Simulation Results

In what follows, we will provide some results obtained from our simulation experiments on collectively pushing cylindrical and cubic boxes with a group of three robots. The specific parameters set in the experiments are given in Table 11.1.

Parameter	Symbol	Value
number of robots	M	3
number of contact segments	$\mathcal{N}$	32
environment size		100×100
translation step coefficient	α_1	0.5
rotation step coefficient	α_2	$\frac{\pi}{6}$
radius of a cylindrical box		2
dimension of a rectangular-shaped robot		1×1.5
chromosome length	L	30
population size	$\mathcal{P}$	8
crossover probability	p_c	0.8
mutation probability	p_m	0.05

TABLE 11.1. Parameters as used in the experiments.

Case 1: Pushing a Cylindrical Box

Figures 11.14 and 11.15 present a series of snapshots showing the movement steps of three robots in pushing a box toward a desired goal location. Figures 11.16 and 11.17 show the trajectories of the robots and the box. In the figure, the solid line denotes the trajectory of the box; and the dashed line, dotted line and dash-dot line correspond to those of the three robots. Figure 11.18 shows the corresponding fitness value changes during group behavior learning.

At the beginning of box-pushing, group robots move in a rather chaotic manner. The reason is that when the chromosomes in a population are randomly initialized, no high-fitness configuration can be found with genetic operations. It takes more than 30 generations for the group of distributed robots to emerge more effective group behaviors. This is also reflected in the fitness value changes as shown in Figure 11.18. The motion trajectories are getting increasingly smoother, as the evolution for collective box-pushing continues. After about 40 generations, as shown in the figures, the coordination among the robots becomes most apparent.

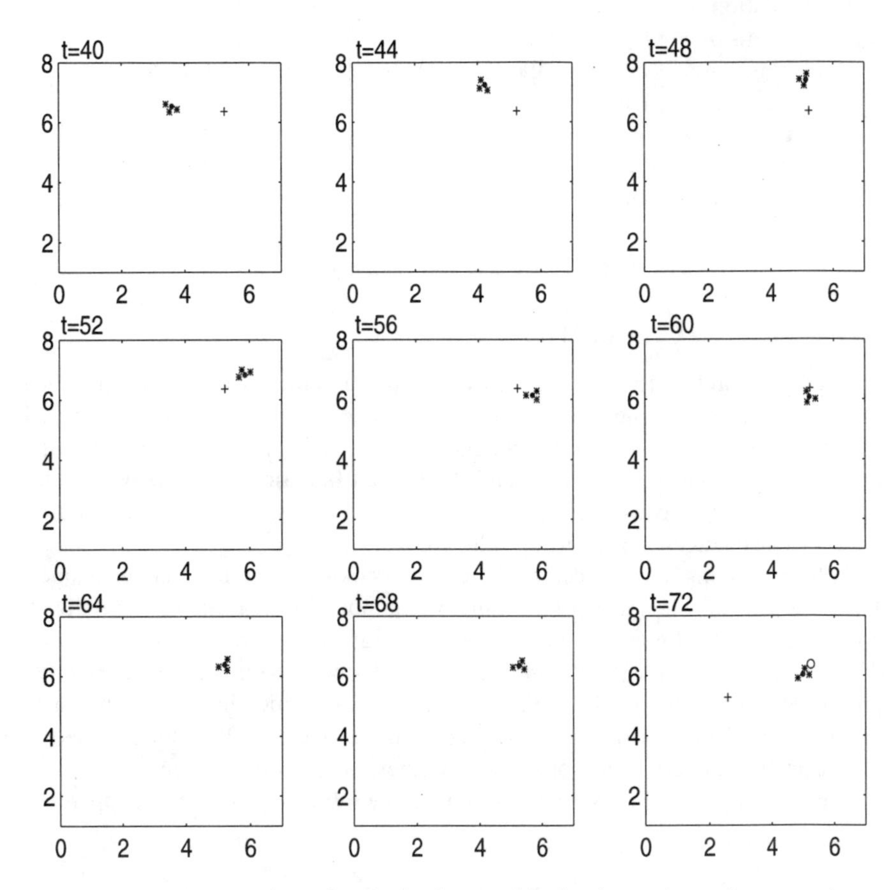

FIGURE 11.7. The snapshots of collective box-pushing by three robots taken every 4 generations from generations 40 to 72 (top view). The collective movement in each snapshot is generated based on *the fittest member of a population* at a certain generation of collective behavior evolution. In the robot environment, a desired goal location is marked with symbol +. A box, marked with •, is gradually moved toward the goal direction, after about 40 generations of motion strategy selection (©1999 IEEE).

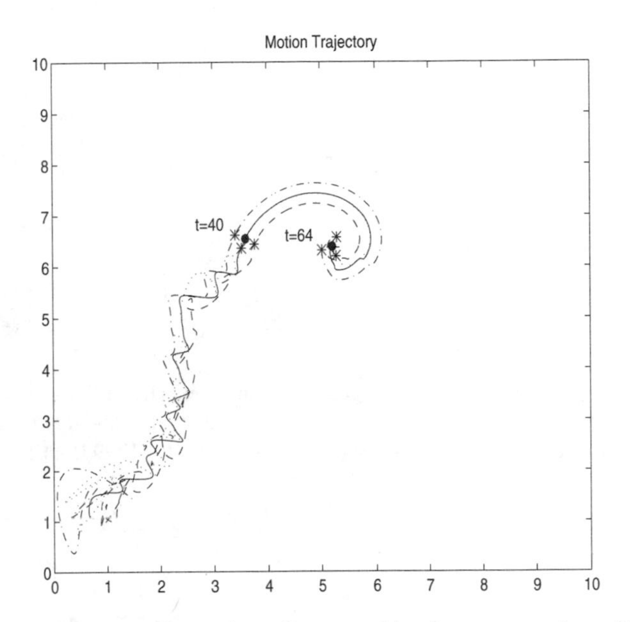

FIGURE 11.8. The box-pushing trajectories created by three group robots. The solid line corresponds to the trajectory of the box, whereas the others correspond to the movement traces for the three robots. At the beginning, the net pushing force of the robots results in a rather randomized motion of the box. This corresponds to the random initialization of potential collective motion strategies in the genetic algorithm. Next, the motion of the box is transformed into a phase of chaotic oscillation, as the genetic algorithm is undergoing the recombination and mutation of some suboptimal local movement strategies (that is, certain terms in the fitness function may first get optimized before the others). Later, after some generations of selection, a niche is found, representing a globally near-optimal collective motion strategy, and hence the oscillating movement of the box becomes stabilized (©1999 IEEE).

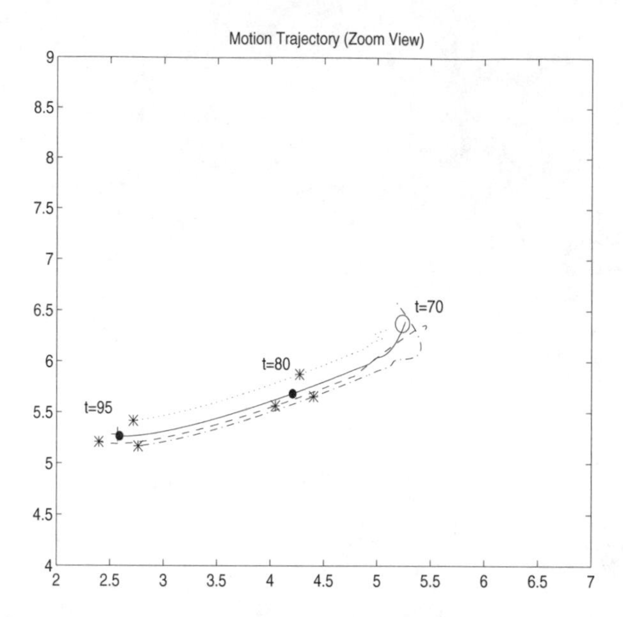

FIGURE 11.9. At generation 70, a new goal location is set. The earlier selected movement strategies for group robots to collectively push a box can be immediately readjusted, since only the movement direction needs to be offset in this case (©1999 IEEE).

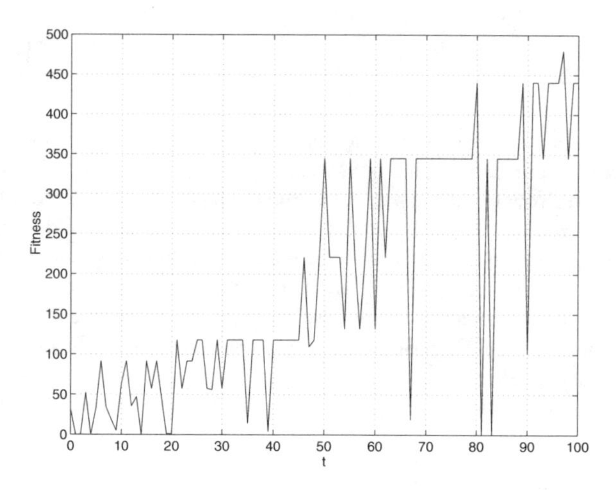

FIGURE 11.10. A fitness function plot for *the fittest member of a population* at each generation, over 100 generations (the entire evolution process of collective box-pushing). Due to the fact that any slight movement change can cause significant variations in some of the terms in the fitness function (the direction and/or the magnitude of the net pushing force), the fitness values recorded have shown some fluctuations. Nevertheless, the trend of the fitness change tends to increase over the period of collective behavior evolution, even when the desired goal locations are reset twice at generations 70 and 97 (©1999 IEEE).

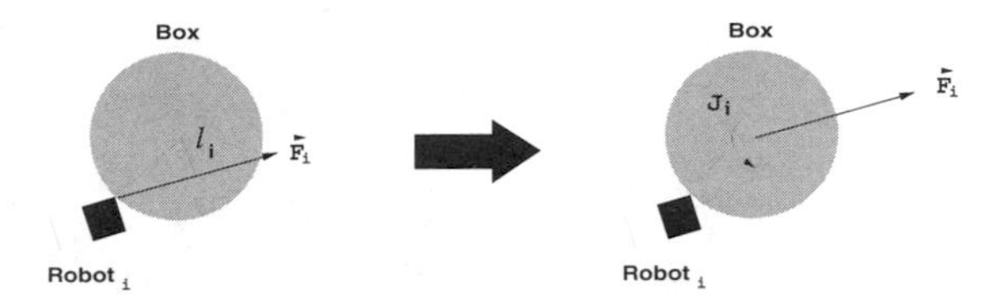

FIGURE 11.11. A top view of group robots pushing a cylindrical box (Note: for the sake of clarity, only one of the three robots is shown). The pushing force on the box is created by individual, rectangular-shaped robots via their contact points. As a result of this force, the box may be pushed to translate as well as rotate about a certain axis.

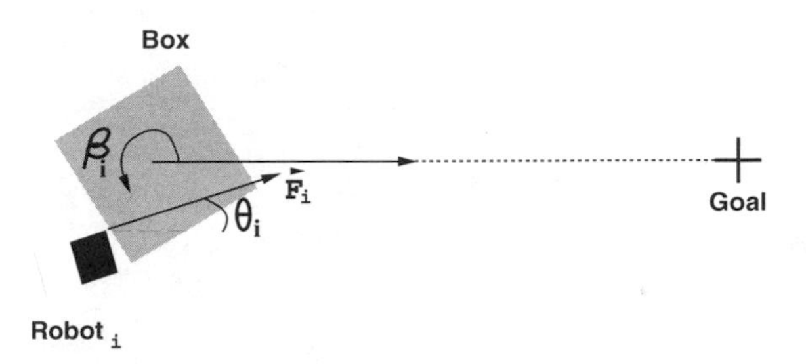

FIGURE 11.12. A top view of group robots pushing a cubic box (Note: for the sake of clarity, only one of the three robots is shown). As in the preceding case, the pushing force on the box is created via the physical contacts between the robots and the box. During collective box-pushing, the relative spatial configuration of each robot is specified by a relative orientation, β_i, with respect to a desired goal location and a reference point on the box.

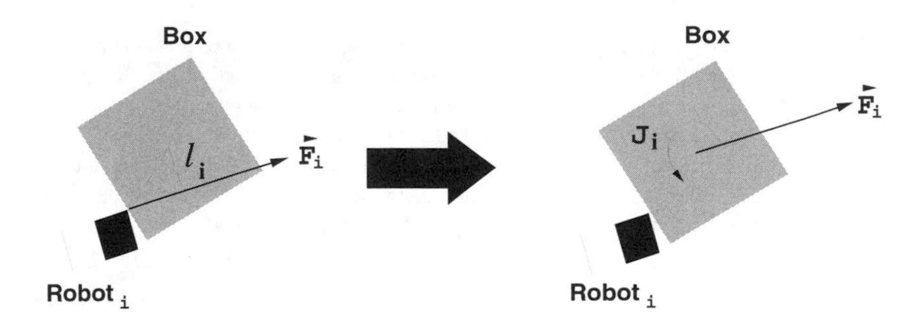

FIGURE 11.13. As a result of the net contact force on a box, as created by three group robots, the box may be pushed to translate as well as rotate about a certain axis (Note: for the sake of clarity, only one of the three robots is shown).

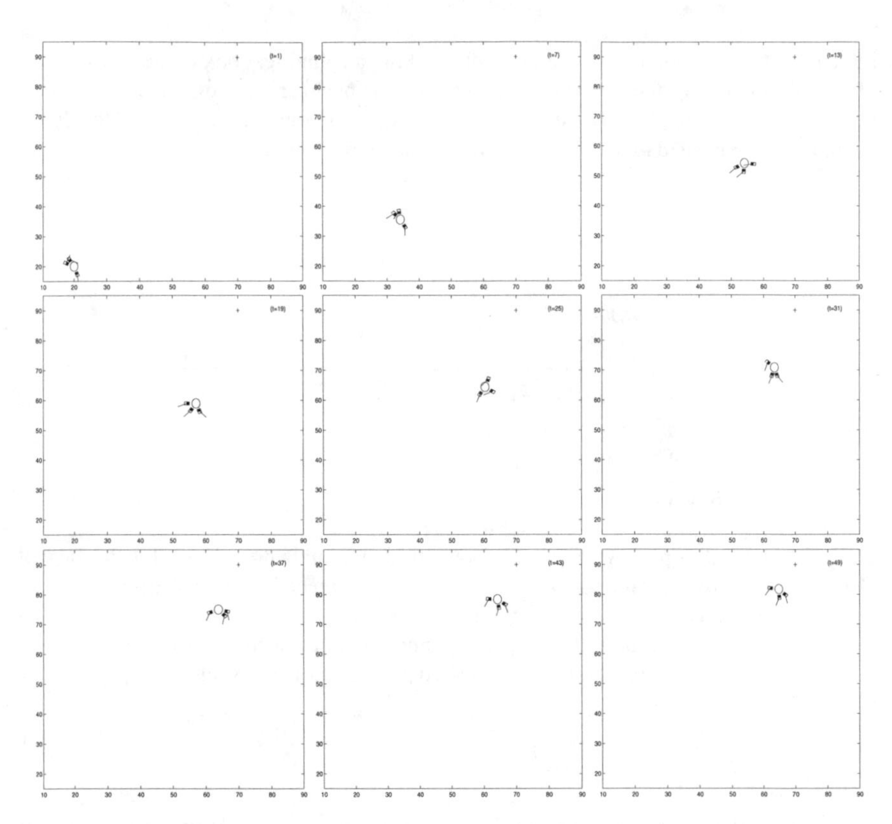

FIGURE 11.14. The snapshots of pushing a cylindrical box by three group robots (top view). The collective movement in each snapshot is generated based on *the fittest member of a population* at a certain generation of collective behavior evolution. In the robot environment, a desired goal location is marked with symbol +. The number shown at the upper right corner of each snapshot corresponds to the generation number. The tail originated from each robot indicates the direction of its movement. As shown in the figure, good collective motion strategies are found after about 30 generations.

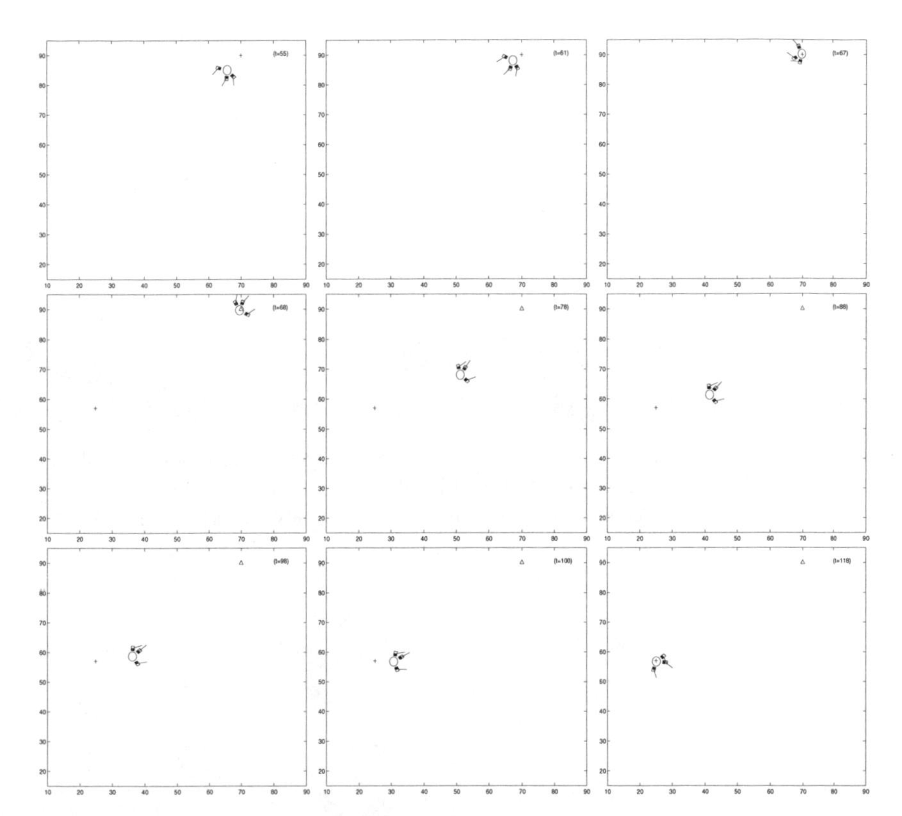

FIGURE 11.15. The snapshots of pushing a cylindrical box by three group robots (top view). The collective movement in each snapshot is generated based on *the fittest member of a population* at a certain generation of collective behavior evolution. In the robot environment, a desired goal location is marked with symbol +. The number shown at the upper right corner of each snapshot corresponds to the generation number. The tail originated from each robot indicates the direction of its movement. As shown in the figure, at generation 67, the desired goal location is reset. In such a case, previously found motion strategies are quickly reselected.

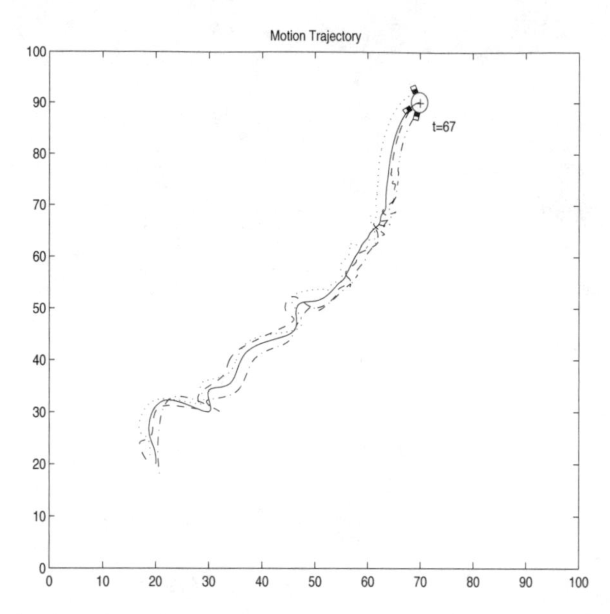

FIGURE 11.16. The trajectories of cylindrical box-pushing by group robots from the beginning until the box reaches a desired goal location. The solid line corresponds to the trajectory of the box, whereas the others correspond to the movement traces for the three robots.

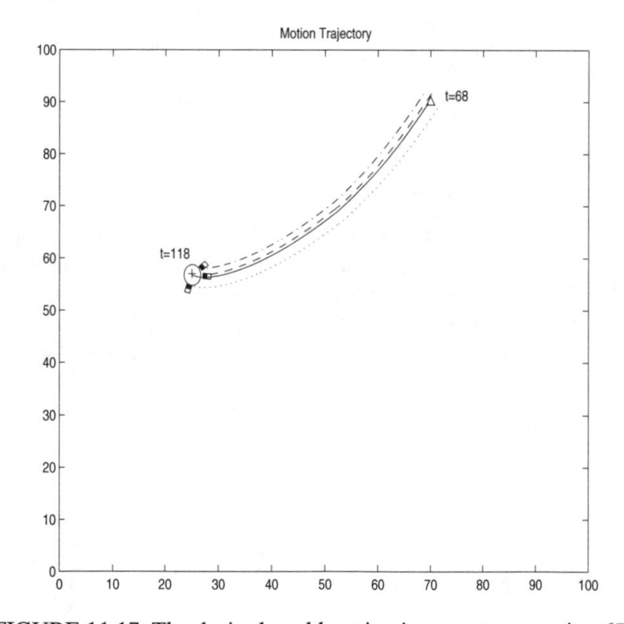

FIGURE 11.17. The desired goal location is reset at generation 67.

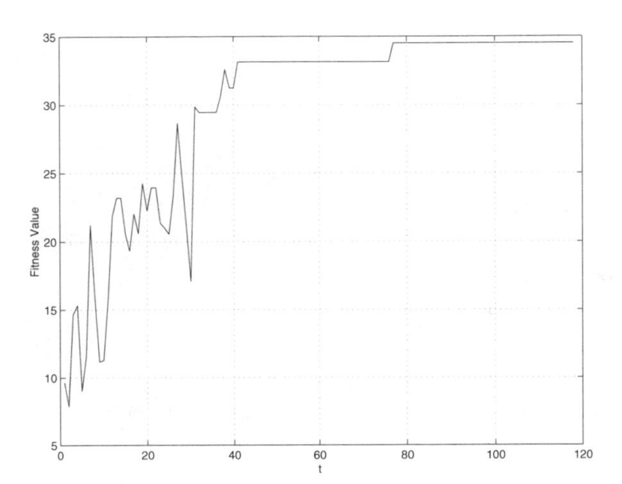

FIGURE 11.18. A fitness function plot for *the fittest member of a population* at each generation over the entire collective behavior evolution of 118 generations. Unlike the previous illustrative example, in this experiment, changes in the movements of three group robots can be quickly stabilized after about 30 generations, where further recombinations and mutations of the potential motion strategies will not result in significant fluctuations in the fitness values. This is primarily because the present fitness function, e.g., the direction and/or the magnitude of the net pushing force, is not as sensitive as the previous one to the same amount of movement variations.

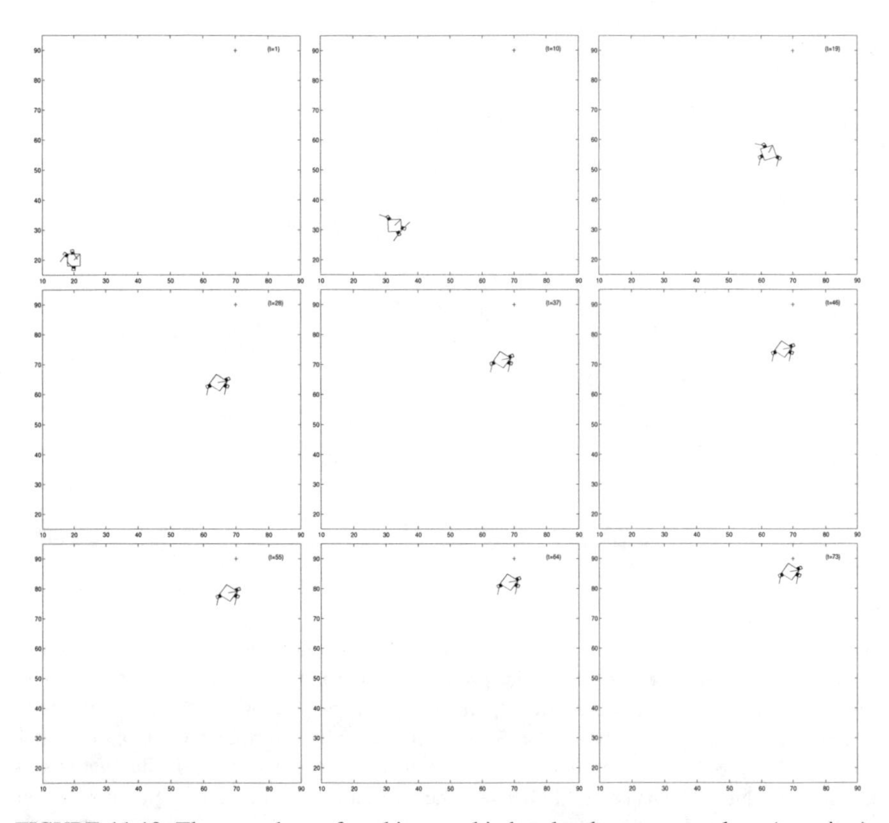

FIGURE 11.19. The snapshots of pushing a cubic box by three group robots (top view). Similar to the preceding experiment, the collective movement in each snapshot is generated based on *the fittest member of a population* at a certain generation of collective behavior evolution. Except the change of a box, all the rest of notations here remain the same as before. As shown in the figure, good collective motion strategies are found after about 20 generations.

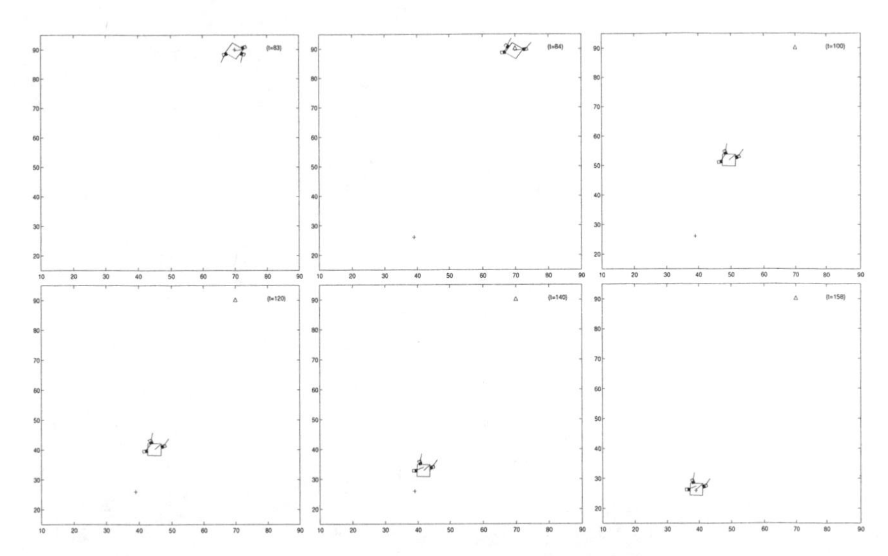

FIGURE 11.20. The snapshots of pushing a cubic box by three group robots (top view). As shown in the figure, at generation 84 the desired goal location is reset. In such a case, previously found motion strategies are readjusted.

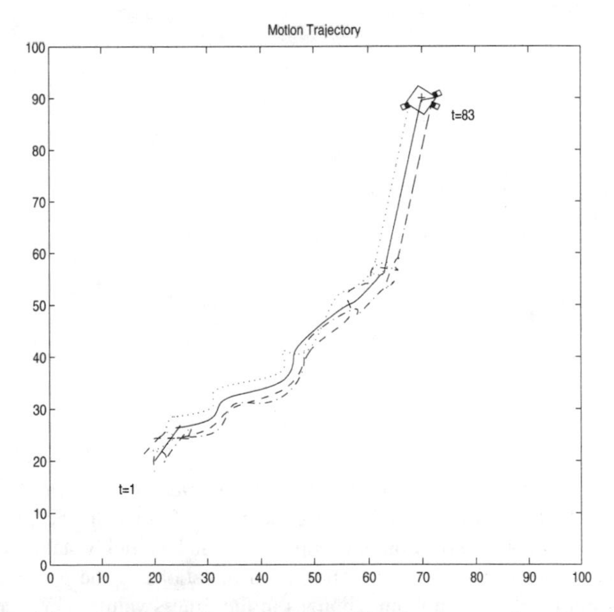

FIGURE 11.21. The trajectories of cubic box-pushing by group robots from the beginning until the box reaches a desired goal location. The solid line corresponds to the trajectory of the box, whereas the others correspond to the movement traces for the three robots.

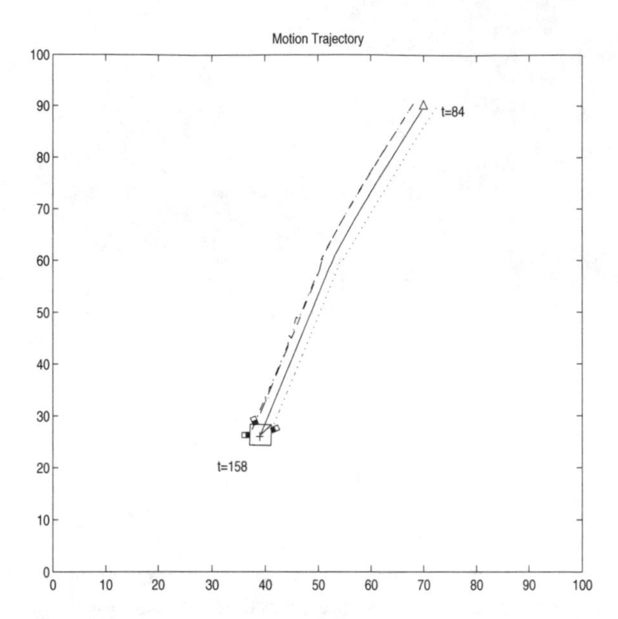

FIGURE 11.22. At generation 84, a new goal location is set.

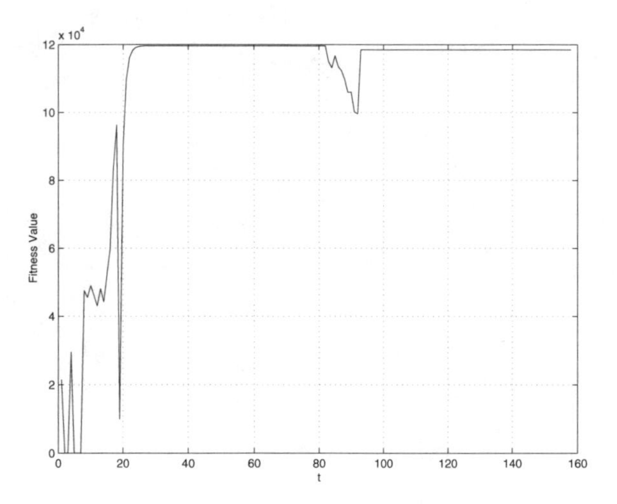

FIGURE 11.23. A fitness function plot for *the fittest member of a population* at each generation over the entire evolution of 158 generations. Similar to the previous experiment, the collective movements of three group robots can be quickly stabilized after about 20 generations, where further recombinations and mutations of the potential movement strategies will not result in significant changes in the fitness values. However, unlike the cylindrical box-pushing experiment, in the present experiment, after the goal location is reset, the robots cannot simply reselect the previously found motion strategies due to the orientation of the box created at the time when the goal is first reached at generation 83.

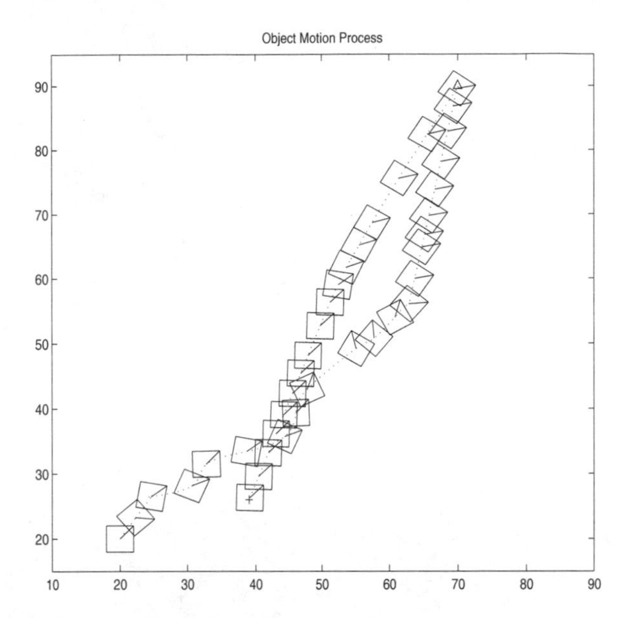

FIGURE 11.24. The trajectory of a cubic box collectively pushed by three group robots from the beginning until the box reaches the first desired goal location, and then moves toward a new goal. The diagonal inside the cubic box signifies the orientation of the box.

Case 2: Pushing a Cubic Box

In the case of cubic box-pushing, all parameters, except the box shape and dimensions, are the same as those in the previous case. The side of the cubic box is 4 units long.

Figures 11.19 and 11.20 show the snapshots of cubic box-pushing. Figures 11.21 and 11.22 give the trajectories traced in the process. Figure 11.23 presents a plot of the corresponding fitness value changes during the behavior evolution.

Similar to the previous experiment, the collective movements of three group robots can be quickly stabilized after about 20 generations, where further recombinations and mutations of the potential movement strategies will not result in significant changes in the fitness values. However, unlike the cylindrical box-pushing experiment, in the present experiment, after the goal location is reset, the robots cannot simply reselect the previously found motion strategies due to the orientation of the box created at the time when the goal is first reached at generation 83.

Figure 11.24 provides a series of box positions and orientations, showing the translation and rotation of the box as a result of collective pushing. We note that the box is being pushed without much rotation (The diagonal inside the box signifies the orientation of the box).

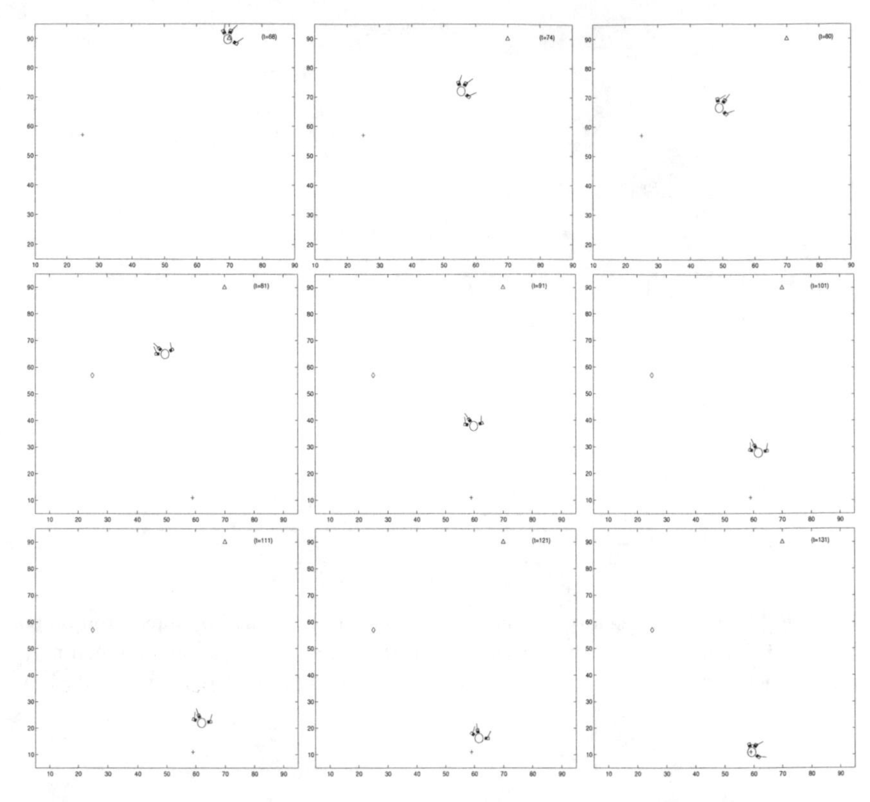

FIGURE 11.25. The snapshots of collective cylindrical box-pushing in which a new goal location, denoted by +, is set at generation 81, before an old one, denoted by ◇, is reached. In this case, the robots successfully readapt their collective movements. The process of collective pushing before generation 68 has been shown in Figures 11.14 and 11.15.

11.4.6.4 Adaptation to Dynamically Changing Goals

In the preceding experiments, a new goal location is set as soon as group robots bring a box to a desired goal location. Here, an interesting question to ask is what will happen if the goal location is changed before it is reached by the box.

Case 1: Pushing a Cylindrical Box

Figure 11.25 shows a series of snapshots from collective cylindrical box-pushing where a new goal location, denoted by +, is set at generation 81, before an old one, denoted by ◇, is reached. In this case, the robots successfully readapt their collective movements. The process of collective pushing before generation 68 has been given in Figures 11.14 and 11.15.

Figure 11.26 gives the trajectories of collective cylindrical box-pushing where the solid line corresponds to the trajectory of the box and the others correspond to those of the three robots. The successful readaptation of the robots is also reflected

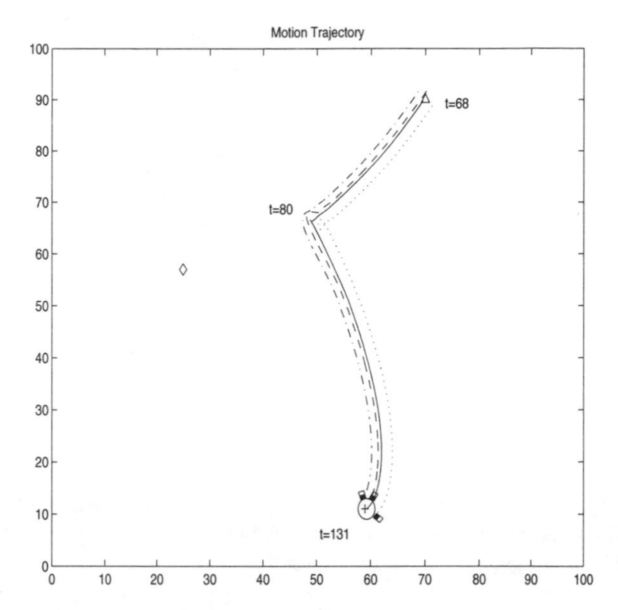

FIGURE 11.26. The trajectories of collective cylindrical box-pushing (the solid line for the box and the others for the three robots). In this case, a new goal location, denoted by +, is set at generation 81, before an old one, denoted by ◇, is reached. The successful readaptation of the robots is reflected in the smoothness of the observed trajectories. The process of collective pushing before generation 68 has been given in Figures 11.16 and 11.17.

in the smoothness of the observed trajectories. The process of collective pushing before generation 68 can be found in Figures 11.16 and 11.17.

Figure 11.27 provides a fitness function plot corresponding to the movement selection and readaptation of three group robots, as shown in Figure 11.26, over the entire evolution process of 131 generations.

Case 2: Pushing a Cubic Box

Figure 11.28 presents several snapshots of collective cubic box-pushing. In this case, a new goal location, denoted by +, is set at generation 91, before an old one, denoted by ◇, is reached. Like the preceding experiment, the robots again successfully readapt their collective movements. The process of collective pushing before generation 84 can be found in Figures 11.19 and 11.20.

Figure 11.29 shows the trajectory of a cubic box collectively pushed by three group robots. The successful readaptation of the robots is reflected in the smoothness of the observed trajectory. In the trace, the diagonal inside the cubic box signifies the orientation of the box.

Figure 11.30 presents a fitness function plot corresponding to the movement selection and readaptation of three group robots, as shown in Figure 11.29, over the entire evolution process of 152 generations.

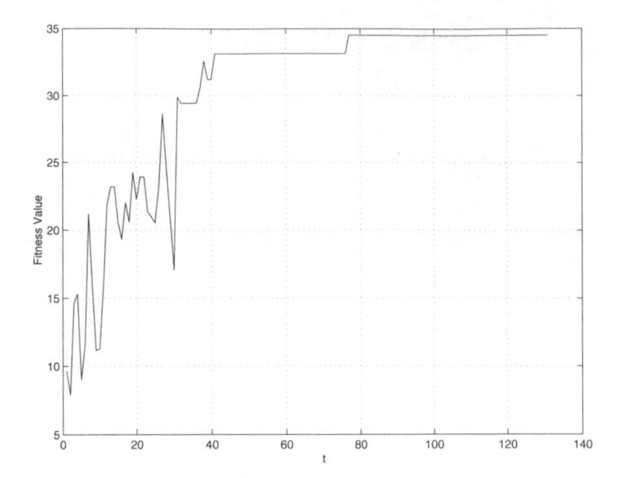

FIGURE 11.27. A fitness function plot corresponding to the movement selection and readaptation of three group robots, as shown in Figure 11.26, over the entire evolution process of 131 generations.

11.4.6.5 Discussions

We can note from the above case studies that group robots can quickly adapt to a new goal location based on their early evolved movement strategies. In the case of cylindrical box-pushing, the adaptation simply involves the reselection of a previous collective movement behavior. This is because the orientation of the cylindrical box is uniform, and its resulting orientation at the time when it reaches an old goal will not affect the subsequent learning to reach a new goal. In fact, such a case may be viewed as if the robots are undertaking a continuous evolution for collective box-pushing toward *the same goal location*.

On the other hand, in the case of cubic box-pushing, the group robots will slightly adjust their motion strategies with respect to the current orientation of the cubic box, in order to move the box toward a new goal location. This is also reflected in the slight perturbation in the fitness function plot of Figure 11.30.

In all experiments presented in this section, the box, either cylindrical or cubic, is pushed by the group robots through their direct contact forces. Since the size of the box is relatively larger than the one used in the preceding illustrative example of Section 11.3, the fitness function that measures the direction and magnitude of the net pushing force/torque will be less sensitive to the slight movement variations as compared to the previous illustrative example.

It should be pointed out that the fitness functions used in the experiments are not unique. Other forms of fitness measurements may also be introduced. It is important to note that the definition of a fitness function can play a role in the resulting performance of collective box-pushing behavior learning.

All case studies that we have described here are taken from the simulation runs that we have conducted and regard as typical. Naturally, this leads us to our next question: from a theoretical point of view, is it possible to estimate the

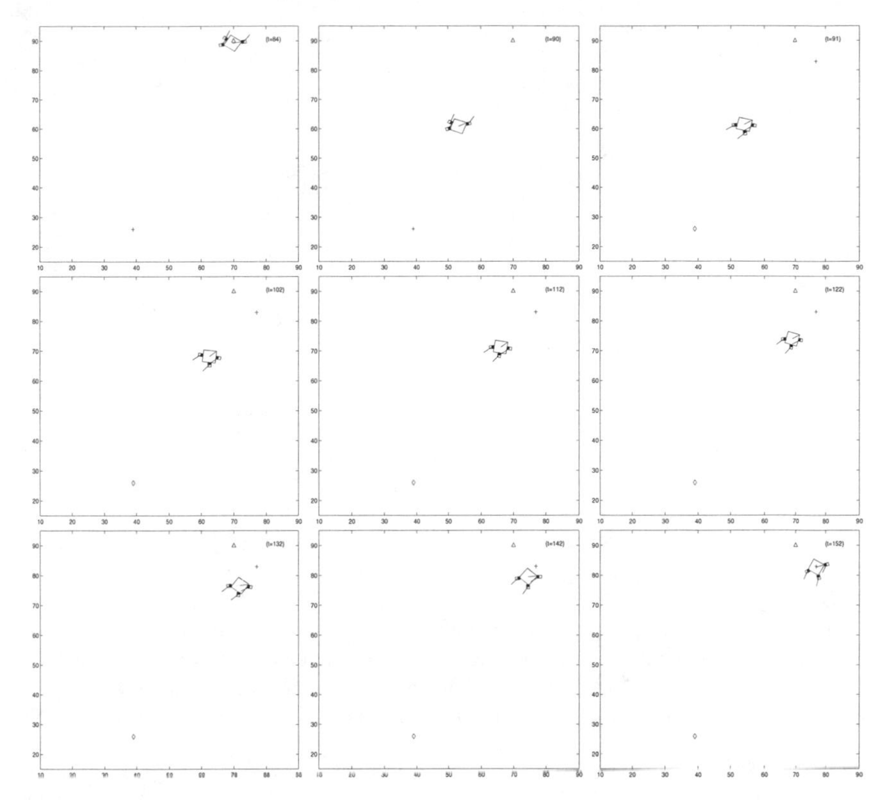

FIGURE 11.28. The snapshots of collective cubic box-pushing in which a new goal location, denoted by +, is set at generation 91, before an old one, denoted by $\diamond$, is reached. Like the preceding experiment, the robots again successfully readapt their collective movements. The process of collective pushing before generation 84 has been shown in Figures 11.19 and 11.20.

probability of converging to a motion strategy by applying the fittest-preserved genetic algorithm? In the next section, we will explicitly deal with this issue.

11.5 Convergence Analysis for the Fittest-Preserved Evolution

In what follows, we will provide some analytical results concerning the convergence of the fittest-preserved genetic algorithm.

11.5.1 The Transition Matrix of a Markov Chain

For a standard genetic algorithm, it is possible to model the set of all possible populations as the state space of a Markov chain and hence derive the transition

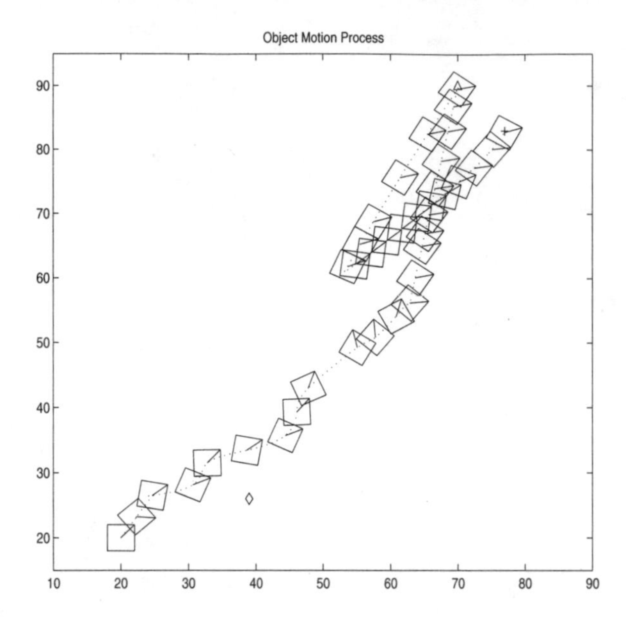

FIGURE 11.29. The trajectory of a cubic box collectively pushed by three group robots. In this case, a new goal location, denoted by +, is set before an old one, denoted by $\diamondsuit$, is reached. The successful readaptation of the robots is reflected in the smoothness of the observed trajectory. In the trace, the diagonal inside the cubic box signifies the orientation of the box.

matrix of the Markov chain. Let $c(0, v)$ denote the occurrences of individual 0 in a population labeled as v, and $c(1, v)$ denote the occurrences of individual 1 in population v and so on. We denote the transition matrix as $Q = (Q_{v,w})$, where $Q_{v,w}$ is the probability that population v evolves to population w and can be calculated as follows [NV92]:

$$Q_{v,w} = \mathcal{P}! \prod_{j=0}^{2^L-1} \frac{b(j,v)^{c(j,w)}}{c(j,w)!}, \tag{11.19}$$

where $\mathcal{P}$, L, and $b(j, v)$ denote the size of a population, the bit length of an individual, and the probability of producing individual j from population v, respectively.

Now let us take a look at the case of fittest-preserved evolution and obtain its corresponding transition matrix. Before doing so, let us consider the following conventions for labeling individuals and populations:

1. We use $\sum_i^L k(i)2^{i-1}$ to label each individual k, where L and $k(i)$ denote the bit length of an individual and the value of the i-th bit of k, respectively. The convention for labeling each individual k becomes: assign labels $j = 0, 1, \cdots, 2^L - 1$ to individuals according to a descending order of their fitness function values, $S_f(k)$. That is, label the individual with the highest

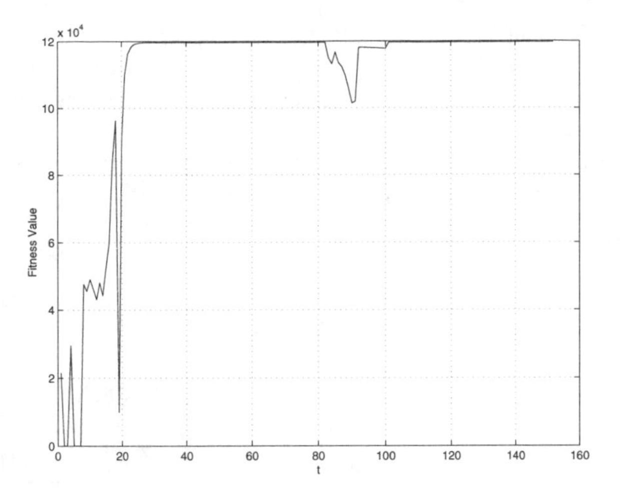

FIGURE 11.30. A fitness function plot corresponding to the movement selection and readaptation of three group robots, as shown in Figure 11.29, over the entire evolution process of 152 generations.

fitness function value 0, label the individual with the second highest fitness function value 1, and so on. Thus, individual k refers to the individual with $(k+1)$-th highest fitness function value. If two individuals have the same fitness function value, they will be differentiated with two distinct labels following certain criteria.

2. The convention for labeling different populations is given as follows: let $k^*(h)$ denote the individual with the highest fitness function value in population h. Assign labels $j = 1, 2, \cdots, N$ to the individuals of each population according to an ascending order of $k^*(h)$. If we find two populations h_1 and h_2 have the same $k^*(h_1) = k^*(h_2)$, we can define some criterion to let them have different labels.

Having given the conventions for labeling individuals and populations, we can now write the transition matrix for the Markov chain that models the fittest-preserved genetic algorithm as follows:

$$Q_{v,w} = \begin{cases} (\mathcal{P}-1)! \prod_{j=0}^{2^L-1} \frac{b(j,v)^{z(j,w)}}{z(j,w)!}, & k^*(v) \geq k^*(w), \\ 0, & k^*(v) < k^*(w), \end{cases} \tag{11.20}$$

where $\mathcal{P}$, L, and $b(j, v)$ denote the size of a population, the bit length of an individual, and the probability of producing individual j from population v, respectively. And,

$$z(j, w) = \begin{cases} c(j, w) - 1, & j = k^*(w), \\ c(j, w), & j \neq k^*(w). \end{cases} \tag{11.21}$$

$Q = (Q_{v,w})$ can be divided into 2^L submatrices along its diagonal. Each submatrix has size $N(i) \times N(i)$, $i = 0, 1, \cdots, 2^L - 1$, where $N(i)$ is the number of populations in which i is the individual with the highest fitness function value, e.g., $k^*(v) = i$.

Specifically, the transition matrix for the fittest-preserved genetic evolution, $Q = (Q_{v,w})$, can be written as follows:

$$\begin{bmatrix} \boxed{Q(0)} & & & & & \\ & \boxed{Q(1)} & & & \mathbf{0} & \\ & & \ddots & & & \\ & & & \overbrace{\boxed{Q(i)}}^{N(i)} \} N(i) & & \\ & \text{nonzero} & & & \ddots & \\ & & & & & \boxed{Q(2^L-1)} \end{bmatrix}. \qquad (11.22)$$

This Markov chain has absorbing states; any initial state will eventually evolve into such states. The absorbing states consist of populations with individual 0.

11.5.2 Characterizing the Transition Matrix Using Eigenvalues

Let $\lambda_1, \lambda_2, \cdots, \lambda_N$ be N eigenvalues of the transition matrix $Q = (Q_{v,w})$. From matrix theory, we know that there exists a matrix, T, such that,

$$Q = \mathrm{T}^{-1} \begin{bmatrix} \lambda_1 & & & \\ & \lambda_2 & & \\ & & \ddots & \\ & & & \lambda_N \end{bmatrix} \mathrm{T}, \qquad (11.23)$$

where

$$\mathrm{T} = \begin{bmatrix} b_{11} & b_{12} & \cdots & b_{1N} \\ b_{21} & b_{22} & \cdots & b_{2N} \\ \cdots & \cdots & \cdots & \cdots \\ b_{N1} & b_{N2} & \cdots & b_{NN} \end{bmatrix}. \qquad (11.24)$$

T^{-1} is the inverse matrix of T, that is,

$$\sum_{l=1}^{N} a_{vl} b_{lw} = \begin{cases} 1, & v = w, \\ 0, & v \neq w, \end{cases} \quad (v, w = 1, 2, \cdots, N). \qquad (11.25)$$

We can obtain

$$Q_{v,w} = \sum_{l=1}^{N} \lambda_l a_{vl} b_{lw}, \tag{11.26}$$

$$Q^n = (Q_{v,w}^{(n)}) = \mathrm{T}^{-1} \begin{bmatrix} \lambda_1^n & & & \\ & \lambda_2^n & & \\ & & \ddots & \\ & & & \lambda_N^n \end{bmatrix} \mathrm{T}, \tag{11.27}$$

and

$$Q_{v,w}^{(n)} = \sum_{l=1}^{N} \lambda_l^n a_{vl} b_{lw}, \tag{11.28}$$

where $Q_{v,w}^{(n)}$ is the probability that population w can be generated from population v after n generations. In our fittest-preserved genetic algorithm, $Q_{v,w}^{(n)}$ can be rewritten in an equivalent form:

$$Q_{v,w}^{(n)} = \sum_{i=0}^{2^L-1} \sum_{j=1}^{N(i)} \lambda_{i,j}^n a_{v,i,j} b_{i,j,w}. \tag{11.29}$$

Let Ω_0 denote the set of populations that include individual 0. Then for $v \in \Omega_0$ and $w \notin \Omega_0$, $Q_{v,w}^{(n)} = 0$. More generally, $Q_{v,w}^{(n)} = 0$ when $k^*(v) < k^*(w)$.

If the initial state is $q^{(0)} = (q_1^{(0)}, q_2^{(0)}, \cdots, q_N^{(0)})$, then the state in the next generation will be $q^{(1)} = (q_1^{(1)}, q_2^{(1)}, \cdots, q_N^{(1)}) = q^{(0)}Q$, and the state after n generations will be $q^{(n)} = (q_1^{(n)}, q_2^{(n)}, \cdots, q_N^{(n)}) = q^{(0)}Q^n$. $\sum_{w\in\Omega_0} q_w^{(n)} = \sum_{w=1}^{N(0)} q_w^{(n)}$ represents the probability that a population including individual 0 will be generated after n generations. On the other hand, $\sum_{w\notin\Omega_0} q_w^{(n)} = 1 - \sum_{w\in\Omega_0} q_w^{(n)}$ represents the probability that no population including individual 0 will be generated. We can show that $\sum_{w\notin\Omega_0} q_w^{(n)}$ is upper bounded.

$$\sum_{w\notin\Omega_0} q_w^{(n)} = \sum_{w\notin\Omega_0} \sum_{v=1}^{N} q_v^{(0)} Q_{v,w}^{(n)}, \tag{11.30}$$

$$= \sum_{w\notin\Omega_0} \sum_{v=1}^{N} \left(q_v^{(0)} \sum_{i=0}^{2^L-1} \sum_{j=1}^{N(i)} \lambda_{i,j}^n a_{v,i,j} b_{i,j,w} \right), \tag{11.31}$$

$$\leq \left(\sum_{w\notin\Omega_0} \sum_{v=1}^{N} \left(q_v^{(0)} \sum_{i=0}^{2^L-1} \sum_{j=1}^{N(i)} a_{v,i,j} b_{i,j,w} \right) \right) |\lambda_{\max}|^n \tag{11.32}$$

$$= \left(\sum_{w\notin\Omega_0} q_v^{(0)} \right) |\lambda_{\max}|^n \quad \text{(derived from Eq. 11.25)}, \tag{11.33}$$

$$= Const\, |\lambda_{\max}|^n, \tag{11.34}$$

where $Const$ is a positive number smaller than 1. Hence, we have:

$$\sum_{w \in \Omega_0} q_w^{(n)} \geq 1 - \; Const \; |\lambda_{\text{max}}|^n \geq 1 - |\lambda_{\text{max}}|^n. \tag{11.35}$$

The next step is to estimate the absolute value of λ_{max}. It is obvious that λ is the eigenvalue of Q if and only if λ is the eigenvalue of one of the submatrices $(Q(i))$ since the determinant of Q is equal to the multiplication of the determinants of $(Q(i))$. Then from Markov chain theory and matrix theory [KS60], we know that

$$|\lambda_{\text{max}}| \leq \mathtt{max}_{0 \leq i < 2^L - 1} \mathtt{max}_{1 \leq j \leq N(i)} \sum_{v=1}^{N(i)} Q(i)_{j,v}, \tag{11.36}$$

where $\sum_{v=1}^{N(i)} Q(i)_{j,v}$ represents the probability that the highest fitness value of population $\sum_{k=0}^{i-1} N(k) + j$ will not get better or worse in the next generation, while $1 - \sum_{v=1}^{N(i)} Q(i)_{j,v}$ represents the probability that the highest fitness value of population $\sum_{k=0}^{i-1} N(k) + j$ will get better in the next generation.

Now let us calculate the least probability that a population will get better in the next generation of box-pushing behavior evolution. From our coding scheme we can prove that, by changing only one bit of the chromosome, it is possible to make the individual corresponding to the chromosome have a higher fitness value. The detailed proof is not given here due to space limitation.

The probability that a mutation happens on the right bit is the least probability that a new population will get better. In fact, the probability can be higher since the crossovers and mutations of other individuals may also generate some individuals with higher fitness values than the original fittest. However, it is sufficient for us to only consider the mutation on the individual with the highest fitness value. When we let the individual with the highest fitness value in a generation get mutated directly, the least probability that a population will get better in the next generation is:

$$(1 - p_m)^{23} \, p_m. \tag{11.37}$$

That is:

$$1 - \mathtt{max}_{0 \leq i < 2^L - 1} \mathtt{max}_{1 \leq j \leq N(i)} \sum_{v=1}^{N(i)} Q(i)_{j,v} \geq (1 - p_m)^{23} \, p_m. \tag{11.38}$$

Thus, we have:

$$|\lambda_{\text{max}}| \leq \mathtt{max}_{0 \leq i < 2^L - 1} \mathtt{max}_{1 \leq j \leq N(i)} \sum_{v=1}^{N(i)} Q(i)_{j,v} \leq 1 - (1 - p_m)^{23} \, p_m. \tag{11.39}$$

Differentiate the right-hand side of the above inequality and let the differentiation be equal to 0, we obtain $(1 - p_m)^{22} (1 - 24 p_m) = 0$. The right-hand side of the inequality will get its minimum value, when $p_m = 1/24$; in such a case,

$|\lambda_{\max}| \leq 1-(1-1/24)^{23} \cdot 1/24 = 0.985$. Now let us revisit Eq. 11.35 and rewrite it as follows:

$$\sum_{w \in \Omega_0} q_w^{(n)} \geq 1 - |\lambda_{\max}|^n \geq 1 - 0.985^n. \tag{11.40}$$

Therefore, we can have:

$$n = 50, \quad \sum_{w \in \Omega_0} q_w^{(n)} \geq 1 - 0.985^{50} \approx 0.53, \tag{11.41}$$

$$n = 80, \quad \sum_{w \in \Omega_0} q_w^{(n)} \geq 1 - 0.985^{80} \approx 0.7, \tag{11.42}$$

$$n = 100, \quad \sum_{w \in \Omega_0} q_w^{(n)} \geq 1 - 0.985^{100} \approx 0.78, \tag{11.43}$$

$$n = 150, \quad \sum_{w \in \Omega_0} q_w^{(n)} \geq 1 - 0.985^{150} \approx 0.9, \tag{11.44}$$

$$n = 200, \quad \sum_{w \in \Omega_0} q_w^{(n)} \geq 1 - 0.985^{200} \approx 0.95. \tag{11.45}$$

We know from the above list that no matter what the initial population is, we can have a probability greater than 50% to get a population containing individual 0 after 50 generations, a probability greater than 78% after 100 generations, and a probability greater than 95% after 200 generations. When we have a population containing individual 0, the direction of the net pushing force by three group robots is exactly toward a desired goal location.

11.6 Summary

In this chapter, we have described a fittest-preserved evolutionary computation approach to learning collective box-pushing behaviors in group robots. We have presented the basic ideas of this approach, its underlying computational architecture, and the fundamentals and characteristics of a genetic algorithm.

We have carried out a series of simulations involving robots and boxes of various sizes and shapes. The experimental results from typical case studies have shown that the coupling of local interaction and global goal-directedness created by the group robots can effectively select a collective box-pushing behavior after a reasonable number of generations.

The above observations are consistent with our theoretical results on the convergence of a fittest-preserved genetic algorithm, as derived from a Markov chain model. Based on Markov chain modeling, we note that there is a high probability for the group robots to optimally achieve a collective box-pushing task.

In order to further examine the adaptation of group robots to the changes in goal locations, we have conducted several case studies where the desired goal locations are reset after and before the box reaches an old goal location. We have noted that in both cases the robots can quickly adjust and sometimes reselect previously found collective motion strategies.

Part IV

Case Studies in Self-Organization

12
Multi-Agent Self-Organization

> *[The man of system] seems to imagine that he can arrange the different members of a great society with as much ease as the hand arranges the different pieces upon a chessboard; he does not consider that the pieces upon the chessboard have no other principle of motion besides that which the hand impresses upon them; but that, in the great chessboard of human society, every single piece has a principle of motion of its own, altogether different from that which the legislator might choose to impress upon it.*[1]
>
> Adam Smith

In this chapter, we will consider the problem of self-organizing a spatial map in an unknown environment by using a group of autonomous robots. Generally speaking, map building in an unknown environment presents a

[1] *The Theory of Moral Sentiments*, Part VI, Chapter 2, Section II.

challenging problem in robotics. Some earlier studies have tackled this problem by using exact search algorithms to derive graph-like representations of the environment. An example of this approach is the work by Betke et al. [BRS94] on the piecemeal learning of a robot environment containing convex obstacles. Others [BBHCD96, VBX96] have addressed the problem by modeling an unknown environment with a set of basic geometric primitives such as line segments, circles, regions, landmarks, and/or local maps. In such an approach, incremental learning algorithms, such as Kohonen neural networks and Kalman filters, are often applied [HBBC96, JGC+97, Koh88, KE94]. While the majority of the map building studies deal with the problems of modeling two-dimensional environments, some researchers [BH94] have investigated the use of a self-organizing approach in reconstructing an unknown three-dimensional surface.

Unlike these studies in which only a single mobile robot or sensing system is involved, this chapter will address the issue of collective artificial potential field map construction using a group of distributed robots.

12.1 Artificial Potential Field (APF)

12.1.1 Motion Planning Based on Artificial Potential Field

The problem of robot motion planning has been traditionally treated as an optimization problem in which the configuration of a robot is represented in a parameter space; and a solution to this problem is computed by searching the parameter space in an attempt to satisfy a predefined cost function, such as the distance between the robot and a goal location. The major limitation of this approach is that it is computationally too costly to generate new plans when dealing with dynamic environments that involve unexpected obstacles. As a more practical approach to real-time planning of collision-free motions for manipulators and mobile robots, the notion of artificial potential field (APF) was proposed by Khatib [Kha85, Kha86]. The APF approach incorporates dynamic sensing feedback into robot control and hence overcomes the limitation by extending the reactiveness of the low-level motion control.

APF theory states that for any goal-directed robot in an environment that contains stationary or dynamically moving obstacles, an APF map can be formulated and computed, taking into account an attractive pole at the goal position of the robot and repulsive surfaces of the obstacles in the environment. This potential field can be expressed as follows:

$$\mathcal{U}_{art}(x) = \mathcal{U}_{goal}(x) + \mathcal{U}_{obs}(x), \tag{12.1}$$

where $\mathcal{U}_{art}(x)$, $\mathcal{U}_{goal}(x)$, and $\mathcal{U}_{obs}(x)$ denote the artificial potential field, the attractive potential from the goal, and the repulsive potential from the obstacles, respectively. x denotes a set of independent parameters, called operational coordinates, that describe the position and orientation of the robot. A possible expression of attractive potential would be:

$$\mathcal{U}_{goal}(x) = -\frac{1}{2}k_p(x - x_{goal})^2, \tag{12.2}$$

where k_p is a positive gain.

An example of repulsive potential is given as follows:

$$\mathcal{U}_{obs}(x) = \begin{cases} \frac{1}{2}\eta\left(\frac{1}{x} - \frac{1}{l_0}\right)^2, & \text{if } x \leq l_0, \\ 0, & \text{if } x > l_0, \end{cases} \tag{12.3}$$

where η is a constant. l_0 is a distance threshold, beyond which no repulsive force will be received by the robot.

Generally speaking, $\mathcal{U}_{obs}$ is chosen such that $\mathcal{U}_{art}$ is a non-negative continuous and differentiable function that tends to infinity when x approaches the surface of an obstacle and tends to zero when x approaches the goal position, x_{goal}. Given Eq. 12.1, the force, resulting from the APF at x, can therefore be derived:

$$\vec{F}_{art} = -\nabla[\mathcal{U}_{art}(x)], \tag{12.4}$$

where ∇ denotes a gradient.

The above expression tells us that applying artificial potential field $\mathcal{U}_{art}(x)$ to a robot can be realized by using $\vec{F}_{art}$ as a command vector to control the robot in its operation space (as the motion of an end-effector can be decoupled in its operation space [Kha87]). In doing so, the joint forces corresponding to $\vec{F}_{art}$ must be obtained using the Jacobian matrix. Under such a control, the robot will be able to avoid obstacles as the repulsive force in the potential field *pushes* it away into the valleys of the field. At the same time, it can move toward a goal location as the attractive force in the potential field *pulls* it in the direction of a global zero-potential pole.

With APF, a robot can reach a stable configuration in its environment by following the negative gradient of its potential field. In this case, locally stable configurations are inevitable. Nevertheless, they can be readily overcome by either incorporating a global motion planner, utilizing a harmonic function that does not contain any local minima, or applying generalized APF formulations, such as *elastic bands* [QK93] and *elastic strips* [BK98]. The generalized APF formulations effectively allow for the real-time planning and control of robot motion that is both locally *reactive* to any dynamically changing obstacles and globally *optimal* with respect to any motion criteria for attaining a predefined goal.

12.1.2 Collective Potential Field Map Building

The question that remains is how the APF theory can be used if a robot environment is not given as *a priori* knowledge. In such a situation, it would be essential to dynamically derive a potential field representation based on the sensory data obtained during the interaction between a robot and its environment.

Prassler [Pra95] has proposed the use of a massively parallel network of simple processing elements, arranged in a rectangular grid structure, for computing and manipulating a two-dimensional potential field. Lee and Kardaras [LK97] have applied a multi-layered backpropagation neural network in building potential field maps of different resolutions. Kassim and Kumar [KK95] have used a single-layered network, called Wave Expansion Neural Network (WENN), for learning artificial potential field maps with respect to the configuration-space (C-space) obstacles of a robot [Lat91].

One of the practical concerns in deriving APF representations with a machine learning technique is that it requires a large amount of input data. As an alternative, we are interested in whether the task of building a potential field representation of an unknown robot environment can be carried out by a group of autonomous robots. Our goal is to enable group robots to collectively perform the map building task based on less sensory data. That is to construct a potential field map as efficiently as possible.

12.2 Overview of Self-Organization

The basic idea behind collective potential field map building is that we utilize a distance association scheme to estimate, from the proximity measurement at one location, the distances between obstacles and other unvisited locations within a neighboring region. Therefore, based on the results of such associations, we can update potential field values corresponding to the measurement at the current location as well as those corresponding to neighboring regions. This association scheme enables a global potential map to be dynamically and incrementally self-organized based on the sensory measurements and associations made by a group of distributed robots. All the robots will share and update the global potential map whenever they have sensory data available. Figure 12.1 presents a schematic flowchart of the process.

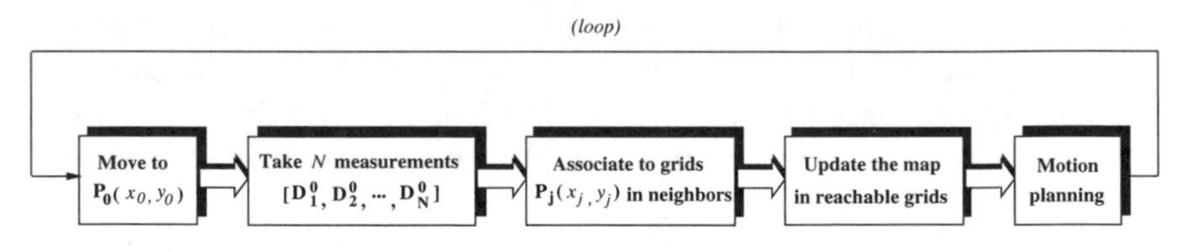

FIGURE 12.1. A flowchart of the map building process.

The system for the collective map building task is illustrated in Figure 12.2. During their interactions with an unknown environment, group robots will broadcast their current locations to a remote host, which is responsible for computing a global potential field map based on the spatial measurements and associations from the robots.

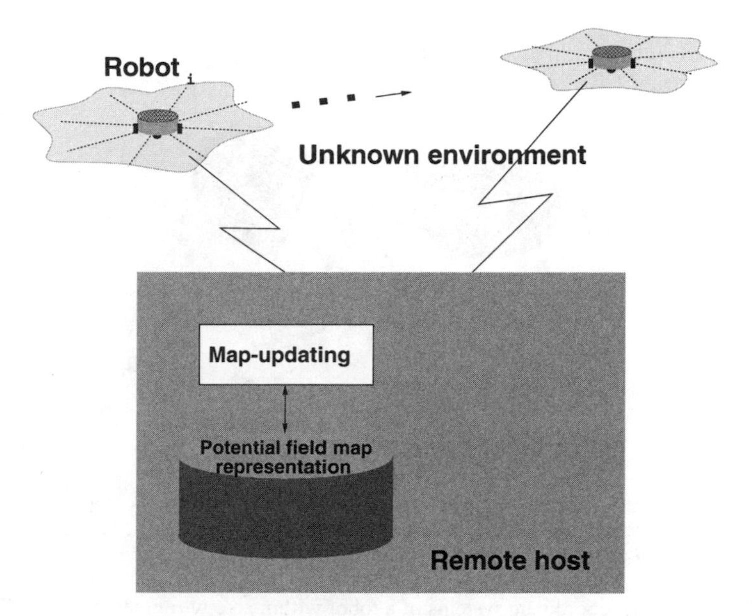

FIGURE 12.2. An illustration of potential field self-organization with a group of autonomous robots.

12.3 Self-Organization of a Potential Field Map

12.3.1 Coordinate Systems for a Robot

The position of an individual robot in a task environment is described using Cartesian coordinates $P_0(x_0, y_0)$. With respect to the current position of the robot, a relative polar coordinate frame can be constructed. Figure 12.3 shows the relationship between the Cartesian coordinates of the robot and its relative polar coordinates. From the figure, it is clear that a new location, $P_j(x_j, y_j)$, can be expressed with respect to $P_0(x_0, y_0)$ as follows:

$$\begin{bmatrix} x_j \\ y_j \end{bmatrix} = \begin{bmatrix} x_0 \\ y_0 \end{bmatrix} + \rho_j \begin{bmatrix} cos\gamma \\ sin\gamma \end{bmatrix}, \tag{12.5}$$

where γ and ρ_j denote the relative polar angle and polar radius of location P_j, respectively.

We assume that the robot is capable of measuring and hence determining its current coordinates in the environment.

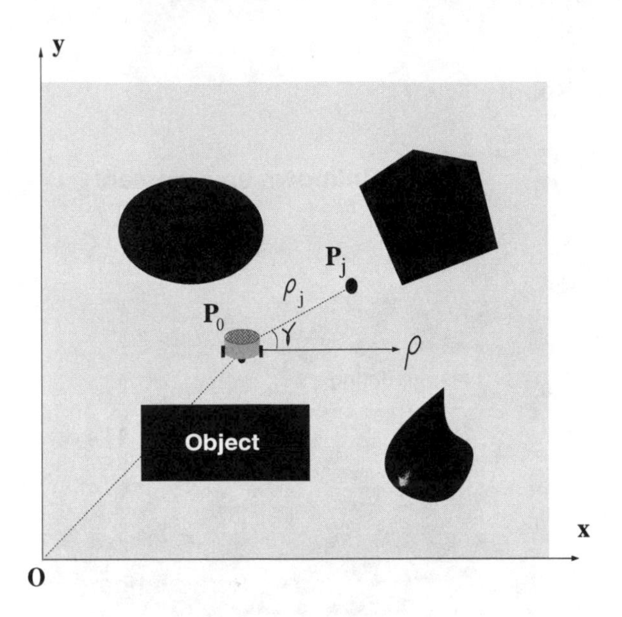

FIGURE 12.3. The coordinate systems of a robot within an environment (©1999 IEEE).

12.3.2 Proximity Measurements

Each robot will measure its distances to the surrounding obstacles in its environment by taking a sensory reading in each of $\mathcal{N}$ directions (with a resolution of $\frac{2\pi}{\mathcal{N}}$ per reading). These measurements will be recorded in a sensing vector, $\mathcal{I}_0$, with respect to location P_0, that is:

$$\mathcal{I}_0 \triangleq \left[D_1^0,\ D_2^0,\ \cdots,\ D_i^0,\ \cdots,\ D_{\mathcal{N}}^0\right]. \tag{12.6}$$

12.3.3 Distance Association in a Neighboring Region

Having recorded the proximity information at P_0 with vector $\mathcal{I}_0$, a robot will associate this information to the neighbor adjacent to P_0 by estimating the proximity values for those locations. The estimated proximity for a neighboring location, P_j, to a sensed obstacle can be determined as follows:

$$\hat{D}_i^j = D_i^0 - \rho_j \cdot cos\beta \ , i = 1, 2, \cdots, \mathcal{N}, \tag{12.7}$$

where $\beta = \gamma_0^{(i)} - \gamma_j$. $\gamma_0^{(i)}$ and γ_j denote the polar angle of the sensing direction and that of location P_j, respectively. $\hat{D}_i^j$ is an estimate for P_j based on the ith direction sensing value. D_i^0 is the current measurement taken from P_0 in the ith direction. Thus, the estimated proximity values for location P_j can be written as follows:

$$\hat{\mathcal{I}}_j \triangleq \left[\hat{D}_1^j,\ \hat{D}_2^j,\ \cdots,\ \hat{D}_i^j,\ \cdots,\ \hat{D}_{\mathcal{N}}^j\right]. \tag{12.8}$$

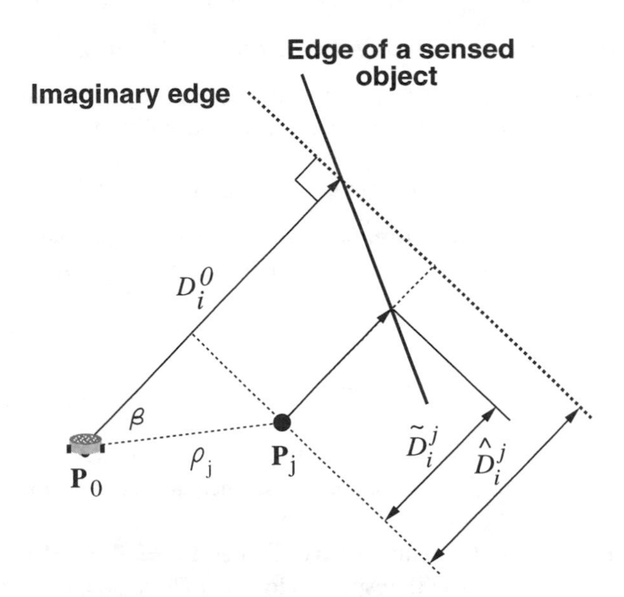

FIGURE 12.4. An illustration of the distance association scheme that computes proximity estimations in a surrounding region based on the proximity measurement obtained at the current location (©1999 IEEE).

Figure 12.4 illustrates the distance association scheme. From the figure, it is clear that an estimated proximity value in one of the $\mathcal{N}$ directions gives true proximity information, if the sensing direction is perpendicular to the edge of a sensed obstacle and the edge is long enough. Otherwise, the estimate provides an approximation; based on Figure 12.4, we can readily derive its error as follows:

$$\varepsilon_d = |\tilde{D}_i^j - \hat{D}_i^j|, \tag{12.9}$$

where $\tilde{D}_i^j$ denotes the true proximity value in the ith sensing direction at location P_j.

It should be pointed out that the amount of errors induced in our association scheme depends on the complexity of a given environment. In order to express our confidence in accepting proximity estimates, we define a confidence weight for each element of $\hat{\mathcal{I}}_j$, that is, a function of the distance between a robot and location P_j, or specifically:

$$w_j = e^{-\eta \rho_j^2}, \tag{12.10}$$

where η is a positive coefficient. ρ_j is the distance as shown in Figure 12.4.

According to Eq. 12.10, the weight is equal to 1 if the robot is located exactly at P_j. That means $\hat{\mathcal{I}}_j|_{j=0}$ is the true value.

12.3.4 Incremental Self-Organization of a Potential Field Map

At time t, the potential field value of a location can be calculated from its $\hat{\mathcal{I}}_j$ vector, as long as the vector satisfies the following condition:

$$\forall i \in [1, \mathcal{N}], \; \hat{\mathcal{D}}_i^j \geq 0. \tag{12.11}$$

Location P_j will be referred to as a *visible* location if the above condition is satisfied. The potential field estimate for visible location P_j will be computed as follows:

$$\hat{\mathcal{U}}_j^t \triangleq \sum_{i=1}^{\mathcal{N}} e^{-\lambda \hat{D}_i^j}, \tag{12.12}$$

where λ is a positive coefficient. If at some locations Eq. 12.11 is not satisfied, the robot will simply discard those locations, since they are considered part of an obstacle.

Thus, at time t, a set of potential field estimates, $\Omega_j^t \triangleq \{\hat{\mathcal{U}}_j^{t_1}, \hat{\mathcal{U}}_j^{t_2}, \cdots, \hat{\mathcal{U}}_j^{t_k}\}$, can be derived by k robots with respect to location P_j; that is,

$$\Omega_j^t = \Omega_j^{t-1} \cup \mathcal{Q}, \tag{12.13}$$

where Ω_j^{t-1} denotes the set of potential field estimates for location P_j at time $t-1$, and

$$\mathcal{Q} = \begin{cases} \hat{\mathcal{U}}_j^{t_k}, & \text{if } \hat{\mathcal{I}}_j \text{ satisfies Eq. 12.11}, \\ \emptyset, & \text{otherwise}, \end{cases} \tag{12.14}$$

where subscript k indicates that the potential value is estimated based on the measurement of the kth robot.

Ω_j^t is associated with a confidence weight set:

$$W_j^t \triangleq \{w_j^{t_1}, w_j^{t_2}, \cdots, w_j^{t_k}\}. \tag{12.15}$$

Hence, at time t, an acceptable potential field value can be readily calculated as follows:

$$\mathcal{U}_j^t = \begin{cases} \hat{\mathcal{U}}_j^{t_i}, & \text{if } \exists i \in [1, k], \; w_j^{t_i} = 1, \\ \sum_{i=1}^{k} \hat{\mathcal{U}}_j^{t_i} \cdot \bar{w}_j^{t_i}, & \text{otherwise}, \end{cases} \tag{12.16}$$

where $\bar{w}_j^{t_i}$ denotes a normalized weight component of W_j^t, i.e.,

$$\bar{w}_j^{t_i} = \frac{w_j^{t_i}}{\sum_{n=1}^{k} w_j^{t_n}}. \tag{12.17}$$

Eqs. 12.13 and 12.16 are referred to as the *incremental self-organization* of a potential field value. Assume that M robots are collectively working in the environment. At time t, after their distributed sensing, association, and potential field self-organization, a global potential field map covering all locations can therefore be obtained.

12.3.5 Robot Motion Selection

Having performed sensing, association, and map updating operations at location P_0, a robot will move to another location to continue the task of map building. In doing so, the robot may apply one of three motion selection mechanisms.

12.3.5.1 `Directional1`

In this motion selection mechanism, the standard deviation of a potential field map for all sensing sectors within a given maximum movement step d_m at time steps t and $t-1$ is calculated. The movement at time $t+1$ will be determined according to (1) motion direction ϕ and (2) motion step d_s. Thus, the location of a robot at time $t+1$, as denoted by P_0^{t+1}, can be written in the following form:

$$P_0^{t+1} = P_0^t + d_s \cdot e^{j\phi}. \tag{12.18}$$

Let array Δ store the standard deviation of the difference between potential field values at time steps t and $t-1$ in each *sensing sector*, whose ith component can be expressed as follows:

$$\Delta_i = \mathtt{std}(\{\iota_{ij} \mid \iota_{ij} = \mathcal{U}_{ij}^{t} - \mathcal{U}_{ij}^{t-1},\ \forall j \in \mathcal{E}_i,\ i = 1, 2, \cdots, \mathcal{N}\}), \tag{12.19}$$

where operator `std` stands for the calculation of standard deviation for a set. $\mathcal{E}_i$ denotes the ith sensing sector.

Let another array Λ stand for the standard deviation of potential field values at time t for all *locations* in the same sensing sector, whose ith component can be written as follows:

$$\Lambda_i = \mathtt{std}(\{v_{ij} \mid v_{ij} = \mathcal{U}_{ij}^{t},\ \forall j \in \mathcal{E}_v,\ |j - i| \leq 1\}), \tag{12.20}$$

where $\mathcal{E}_v$ is determined based on Eq. 12.19.

Thus, in the motion selection mechanism of `Directional1`, the robot will select its movement direction ϕ_i (i.e., in the ith sensing sector) that satisfies:

$$\phi_i|_{\Delta_i = \mathtt{max}(\Delta_1, \Delta_2, \cdots, \Delta_{\mathcal{N}})}, \tag{12.21}$$

and

$$\forall j,\ P_j \notin \mathcal{E}_i, \tag{12.22}$$

where operator `max` returns the maximum value from a set. $\mathcal{E}_i$ denotes the ith sensing sector, and P_j denotes the location of robot j.

After determining the movement sector, the robot will then choose an exact location $P_0^{t+1}(x_0, y_0)$ within the selected sector to go to. The potential field value around the location should satisfy the following condition:

$$(x_0, y_0)|_{\Lambda_i(x_0,y_0) = \mathtt{max}(\Lambda_1,\Lambda_2,\cdots)}. \tag{12.23}$$

12.3.5.2 Directional2

This motion selection mechanism is similar to Directional1 in that a robot selects its movement direction by applying the same strategy as Directional1. However, the next location, $P_0^{t+1}(x_0, y_0)$, within the selected sector is chosen such that it has the minimum standard deviation in the neighboring locations, that is:

$$(x_0, y_0)|_{\Lambda_i(x_0,y_0) = \mathtt{min}(\Lambda_1,\Lambda_2,\cdots)}, \tag{12.24}$$

where operator min returns the minimum value from a set.

12.3.5.3 Random

In this motion selection mechanism, a robot selects its movement direction and movement step randomly, i.e.,

$$\phi_i = \mathtt{rand}([1 \;\; \mathcal{N}]), \tag{12.25}$$

$$d_s = \mathtt{rand}([1 \;\; d_m]), \tag{12.26}$$

where operator rand returns a random number within an interval.

12.4 Experiment 1

12.4.1 *Experimental Design*

The first experiment studies the characteristics of self-organizing potential field map building with autonomous robots applying the above mechanisms. Initially, the robots have a decentralized spatial distribution, as shown in Figure 12.5, where $*$ denotes a robot. There are 4 obstacles in the environment. The parameters as used for this experiment are given in Table 12.1.

A second-moment error for all locations in an obtained potential field map will be calculated as an evaluation of the different motion selection mechanisms. The error is defined as follows:

$$\varepsilon^t \triangleq \sqrt{\frac{1}{K}\sum_{j=1}^{K}(\mathcal{U}_j^t - \bar{\mathcal{U}}_j)^2}, \tag{12.27}$$

where K denotes the total number of locations in the potential field map, and $\mathcal{U}_j^t$ and $\bar{\mathcal{U}}_j$ denote the estimated and the true potential values at location $P_j(x_j, y_j)$, respectively.

The true potential map and its contour map corresponding to the given environment are shown in Figures 12.6(a) and (b), respectively.

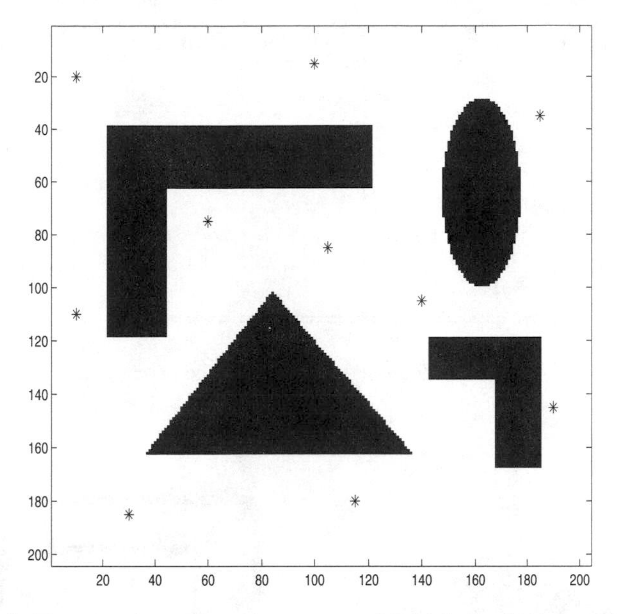

FIGURE 12.5. The experimental environment and initial distribution of the robots in *Experiment 1*.

12.4.2 Experimental Results

Figures 12.7, 12.8, and 12.9 present the potential field maps and their corresponding contour plots at four different time steps with the three motion selection mechanisms, respectively. Figure 12.10 shows the robot trajectories of the past four steps recorded at the four different time steps with respect to the three motion selection mechanisms.

12.5 Experiment 2

12.5.1 Experimental Design

The second experiment examines a scenario in which the group robots that apply the mechanisms have a centralized spatial distribution, as shown in Figure 12.11. The environment is the same as that in *Experiment 1*. The parameters in this experiment are also the same as those given in Table 12.1 except λ. Here, λ is set to $\frac{1}{5}$.

The true potential map and its contour map corresponding to the given environment are shown in Figure 12.12(a) and (b), respectively.

Parameter	Symbol	Unit	Value
number of robots	M		10
sensory section	$\mathcal{N}$		16
environment size		$grid \times grid$	204×204
map resolution		$grid$	5
map locations			40×40
maximum movement step	d_m	$location$	8
coefficient	η		$\frac{1}{600}$
coefficient	λ		1

TABLE 12.1. Parameters as used in the experiments.

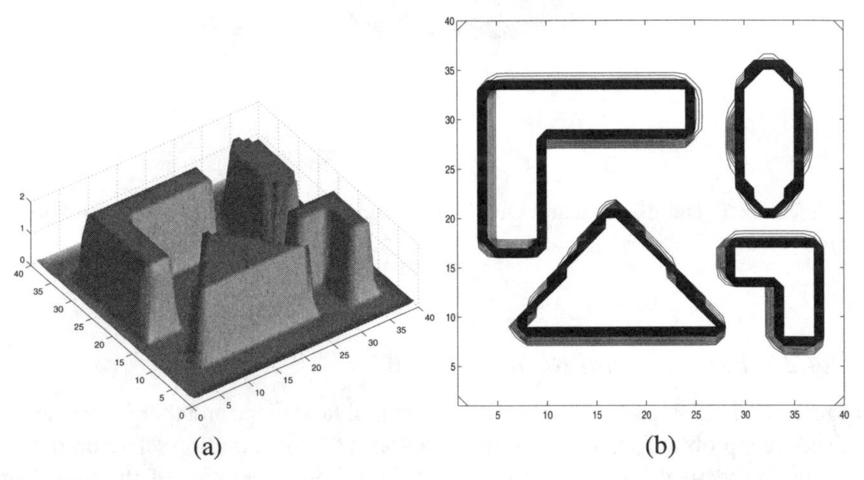

FIGURE 12.6. (a) A true potential field map. (b) A contour plot for the true potential field map.

12.5.2 Experimental Results

Figures 12.13, 12.14, and 12.15 show the potential field maps and their corresponding contour plots at four different time steps with respect to the three motion selection mechanisms, respectively. Figure 12.16 shows the robot trajectories of the past four steps recorded at the four different time steps, as a result of using the three motion selection mechanisms.

12.6 Discussions

From the results provided in Figures 12.7, 12.8, 12.9, 12.13, 12.14, and 12.15, it can be noted that the spatial maps of an unknown environment can be self-organized, which gradually get closer to an exact map. The self-organization method is powerful since only a few measurements from distributed robots are required.

By comparing these figures, it is observed that different motion selection mechanisms have different effects on the collective performance. Clearly, the group robots with the `Random` mechanism is the slowest, especially in the case of a centralized initial distribution as in *Experiment 2*. The other two directional mechanisms are both effective in potential field map building. The reason is that with the directional mechanisms, the robots will be able to explore unknown locations more directly. It should be pointed out that the robots with the `Random` mechanism can still perform the map building task, if they are given enough time. To some extent, this mechanism is effective in a less complex environment.

From the trajectories of group robots with different motion selection mechanisms, observations similar to the above can also be made.

When comparing the results of two directional mechanisms, we can note that `Directional2` performs better at the very beginning of map building. This motion strategy implies that the robots will go to the area within a complex environment but they may stay in the grids with a smooth potential distribution. The robots will go to the undiscovered area at the beginning because it has even features. On the other hand, the robots with `Directional1` will have a relatively small motion area. Thus, `Directional2` enables a faster discovery than `Directional1`. However, after a large area has been discovered, the robots with `Directional1` will become more effective in exploring complex areas, such as the edges of the environment.

When comparing Figure 12.17 with Figure 12.18, it is difficult to identify whether `Directional1` or `Directional2` is better. In Figure 12.17, all three mechanisms have similar error trends. The reason is that because of a decentralized robot distribution, the robots can measure their environment and build a map from various locations. Thus, the merits of directional movement mechanisms cannot be adequately reflected. However, in a centralized initial distribution, we can find that the case of `Random` produces a larger error, and the case of `Directional1` is faster or more accurate than that of `Directional2`. This can be observed from Figure 12.18.

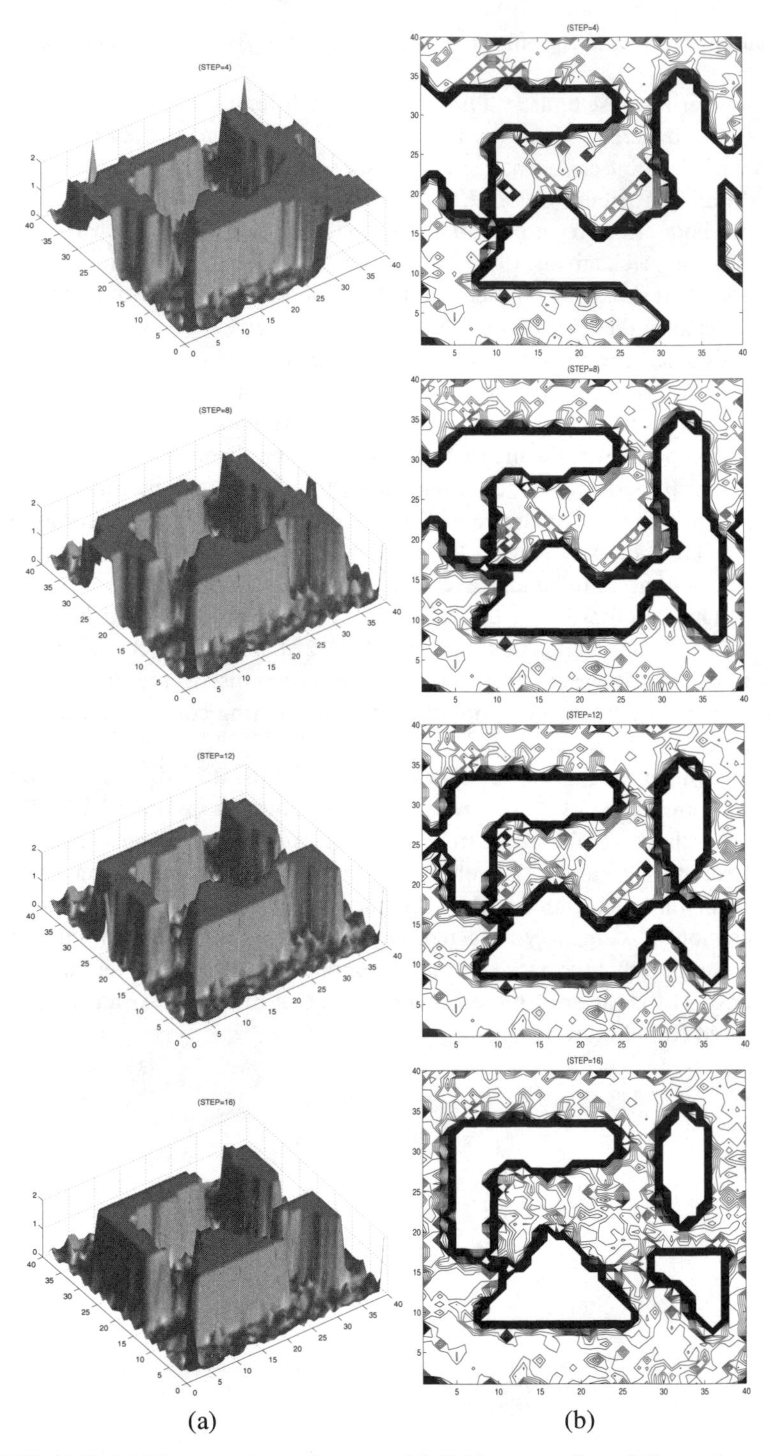

FIGURE 12.7. (a) The snapshots of a potential field map at four different time steps, as built by a group of decentralized robots with the motion selection mechanism of `Directional1`. (b) The corresponding contour plots.

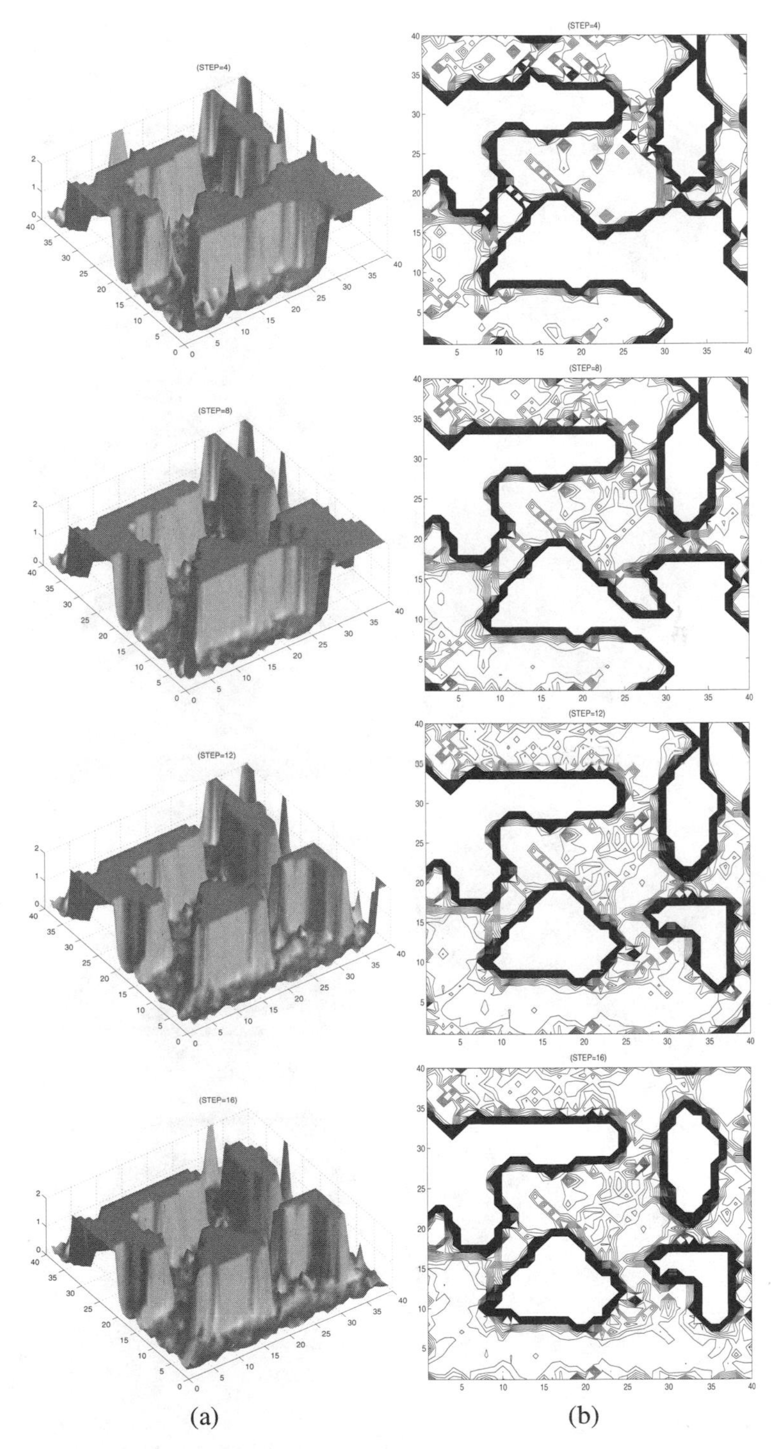

FIGURE 12.8. (a) The snapshots of a potential field map at four different time steps, as built by a group of decentralized robots with the motion selection mechanism of `Directional2`. (b) The corresponding contour plots.

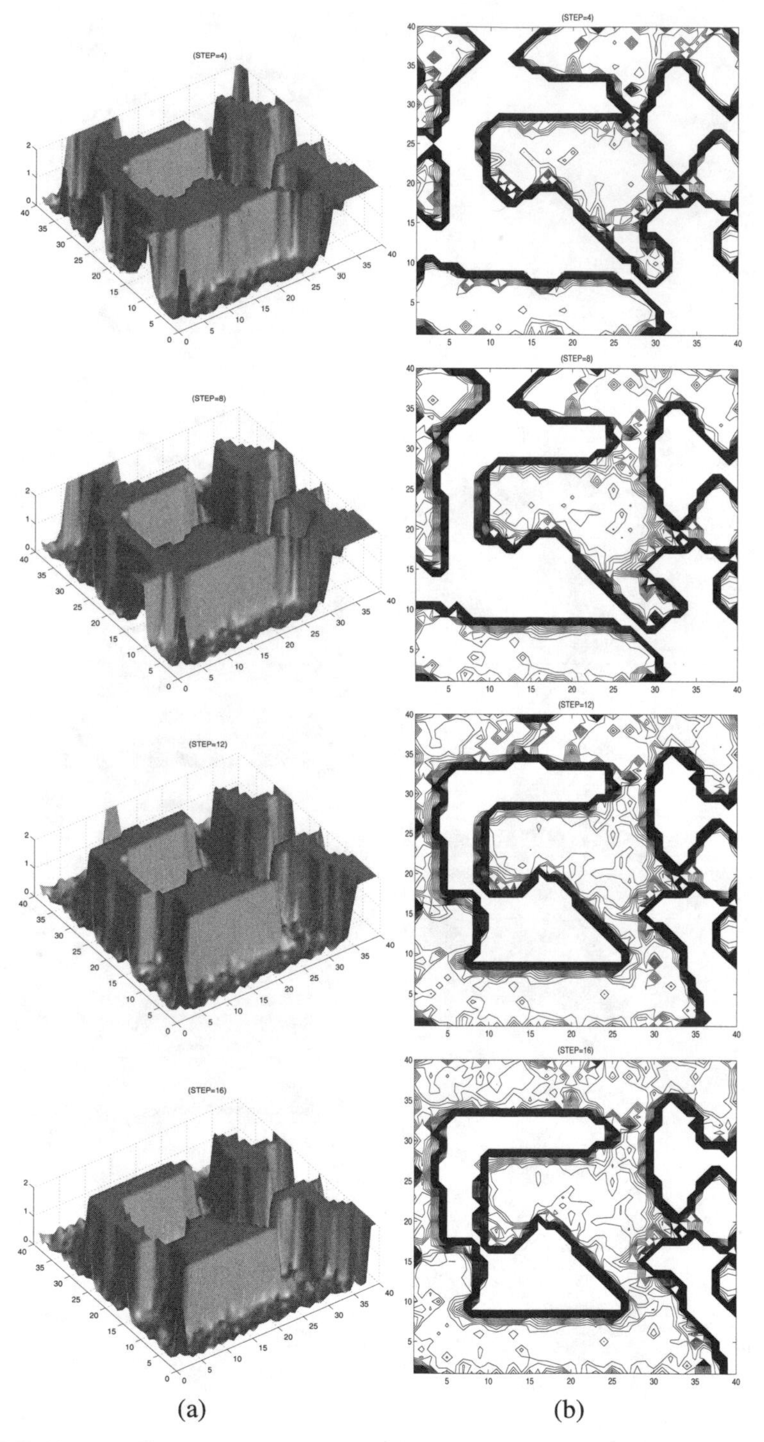

FIGURE 12.9. (a) The snapshots of a potential field map at four different time steps, as built by a group of decentralized robots with the motion selection mechanism of `Random`. (b) The corresponding contour plots.

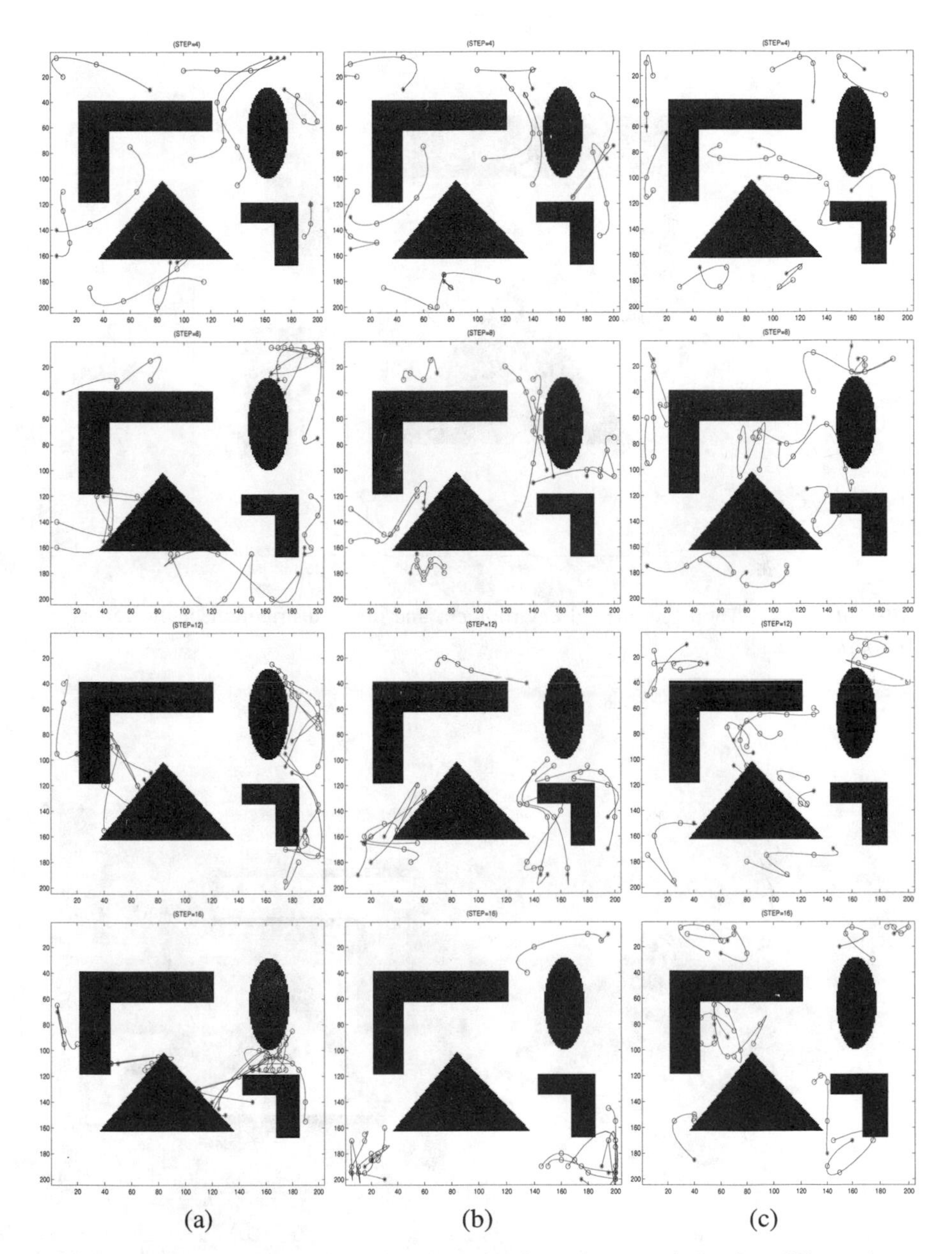

(a) (b) (c)

FIGURE 12.10. The robot trajectories of the past four steps recorded at four different time steps, under the conditions of (a) `Directional1`, (b) `Directional2`, and (c) `Random`.

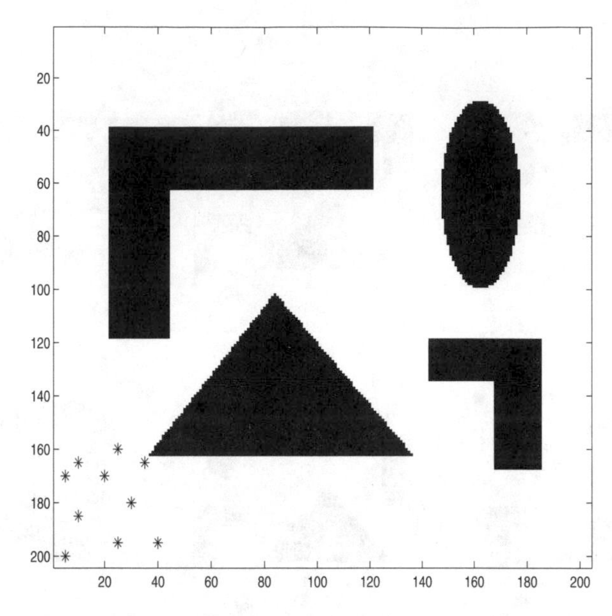

FIGURE 12.11. The experimental environment and initial distribution of the robots in *Experiment 2*.

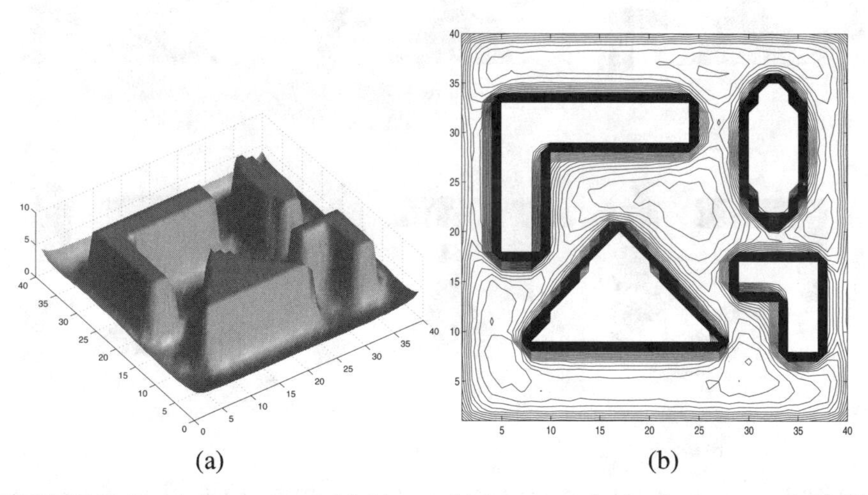

FIGURE 12.12. (a) A true potential field map. (b) A contour plot for the true potential field map (©1999 IEEE).

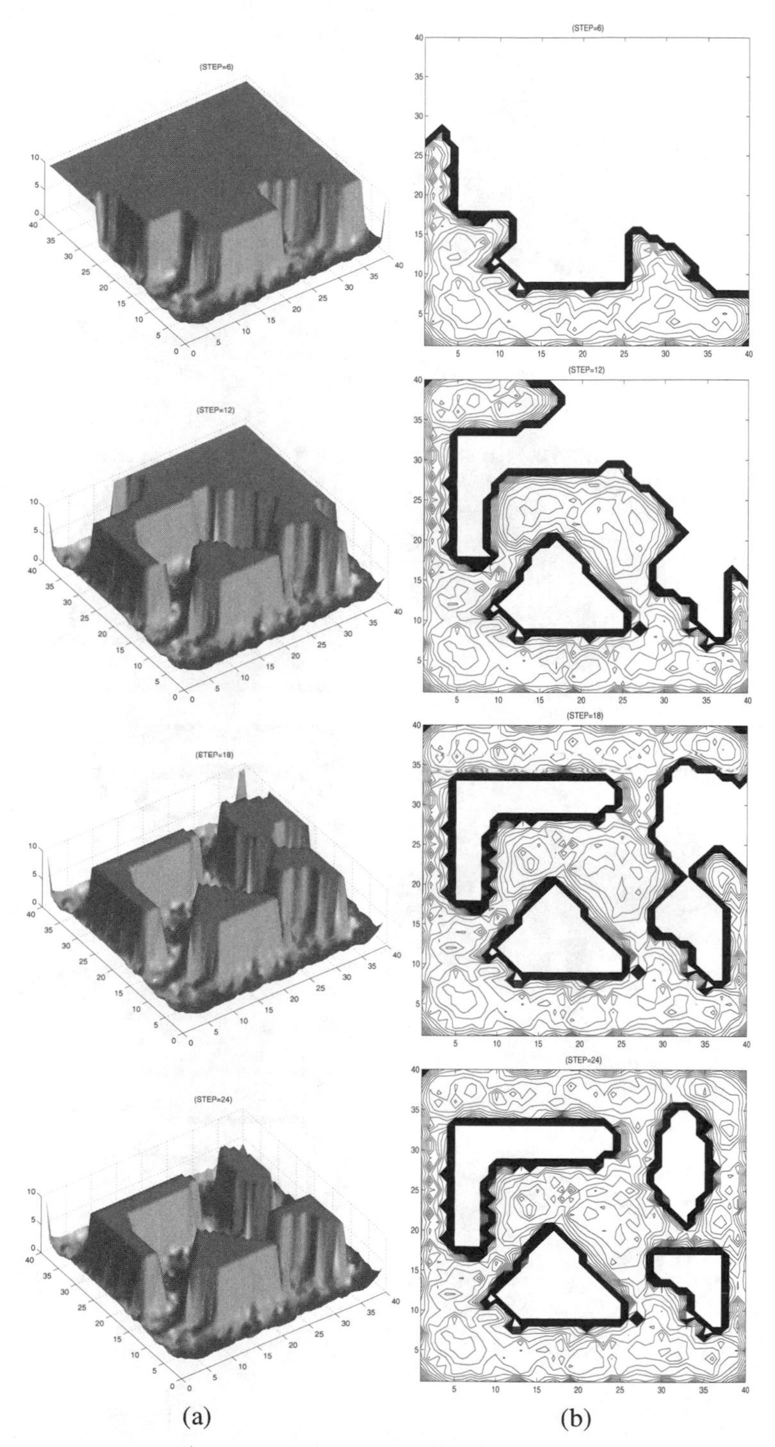

FIGURE 12.13. (a) The snapshots of a potential field map at four different time steps, as built by a group of robots with the motion selection mechanism of `Directional1`. (b) The corresponding contour plots.

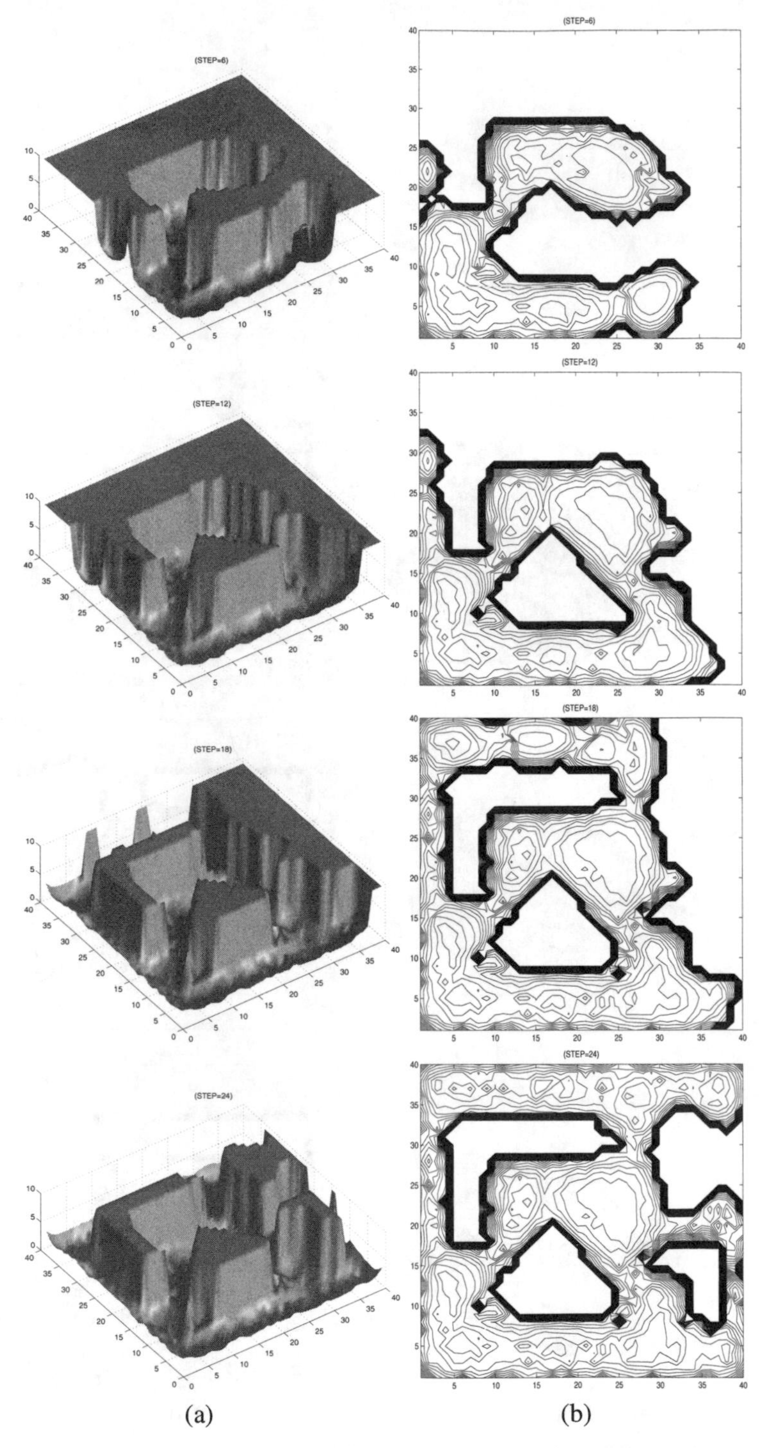

FIGURE 12.14. (a) The snapshots of a potential field map at four different time steps, as built by a group of robots with the motion selection mechanism of `Directional2`. (b) The corresponding contour plots.

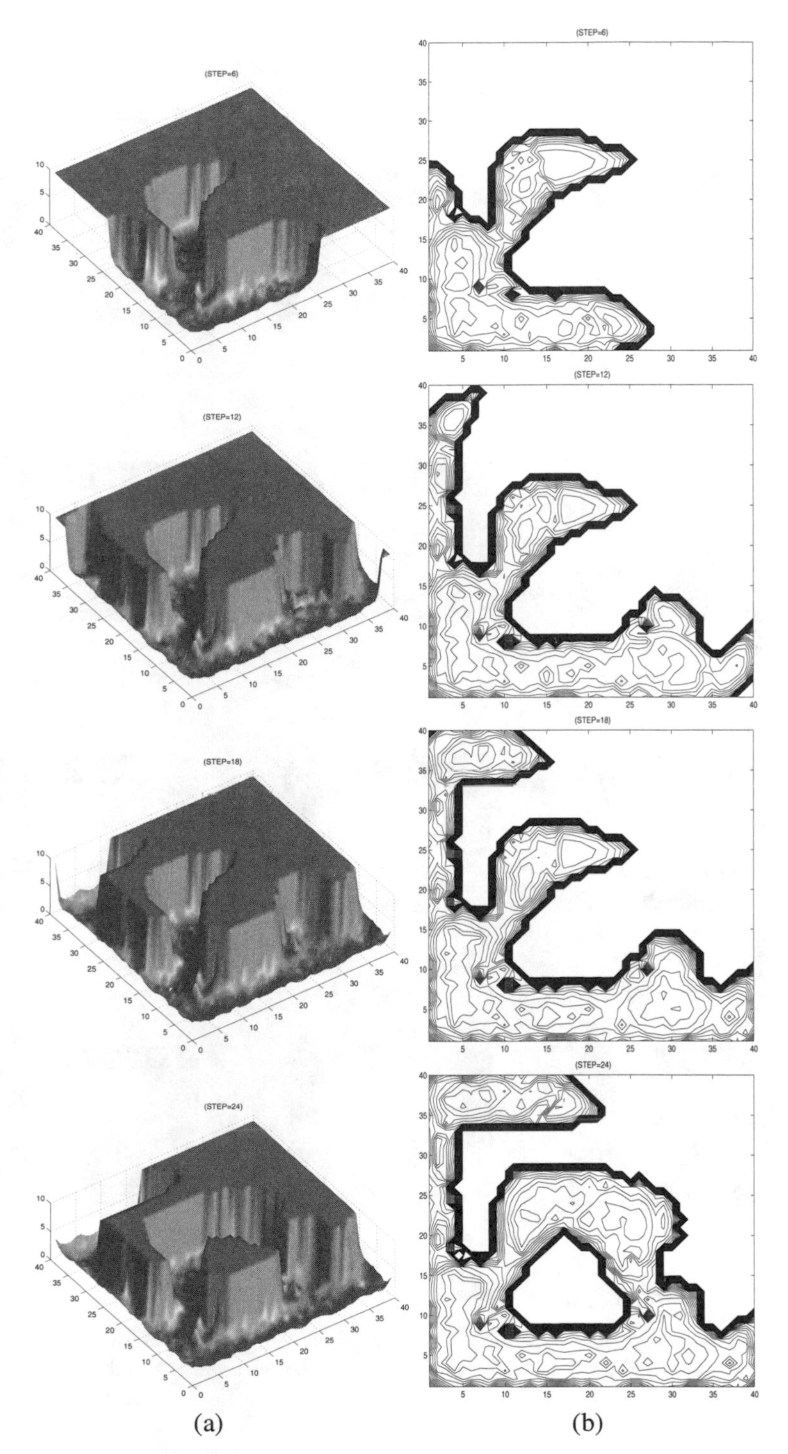

FIGURE 12.15. (a) The snapshots of a potential field map at four different time steps, as built by a group of robots with the motion selection mechanism of `Random`. (b) The corresponding contour plots.

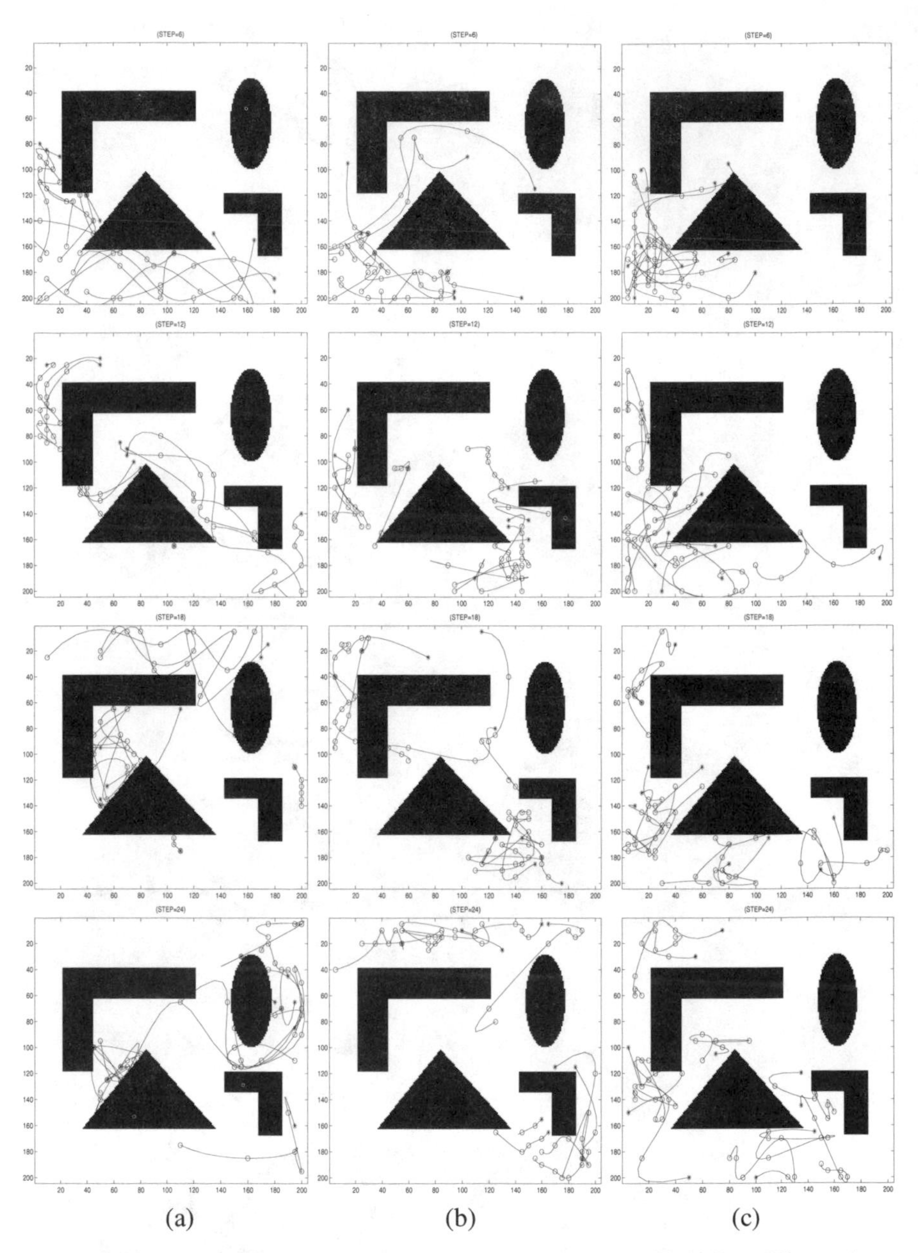

FIGURE 12.16. The robot trajectories of the past four steps recorded at four different time steps, under the conditions of (a) `Directional1`, (b) `Directional2`, and (c) `Random`.

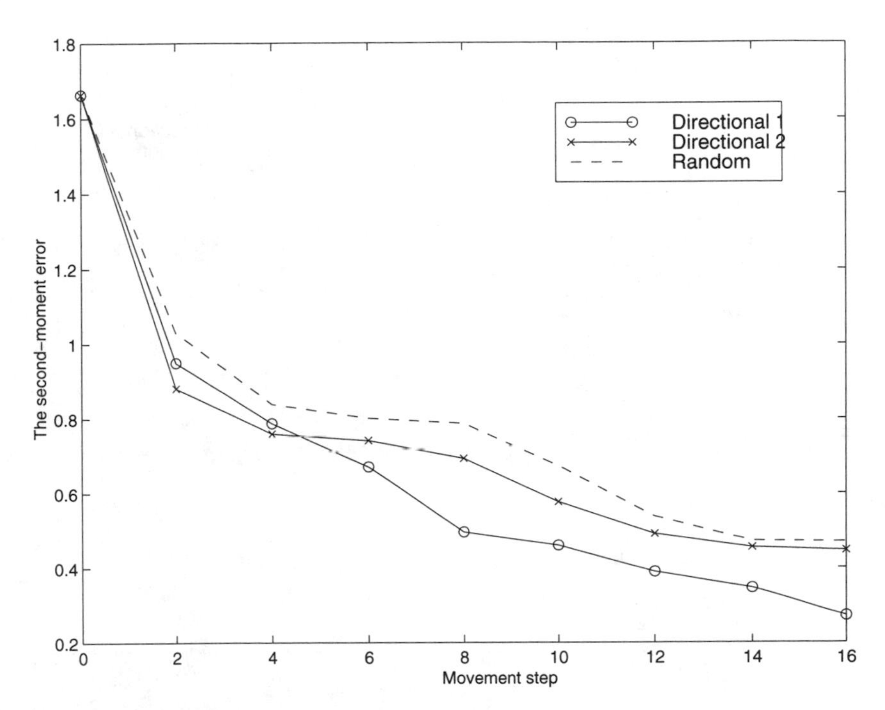

FIGURE 12.17. Second-moment error curves obtained under three different motion selection mechanisms in *Experiment 1*.

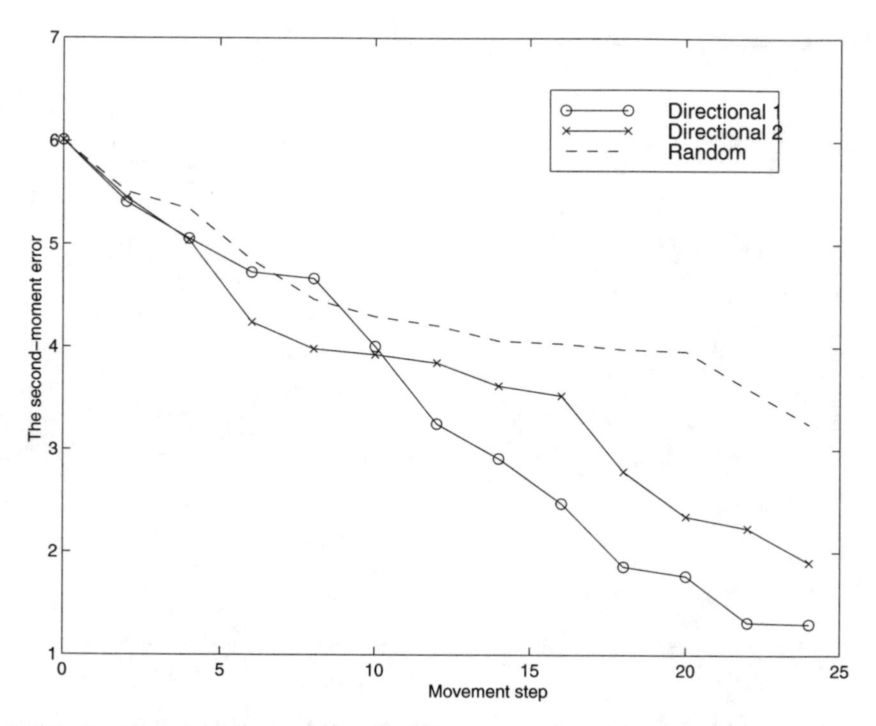

FIGURE 12.18. Second-moment error curves obtained under three different motion selection mechanisms in *Experiment 2*.

13
Evolutionary Multi-Agent Self-Organization

This is what the ecosystem is all about: responsive *mutation. It gives the homostat a task which at last can be done in the time available. The species can evolve, the individual can learn. The viable system, of whatever kind, can adapt. This is a control device which takes the fortuitousness out of randomness.*[1]

Stafford Beer

In Chapter 12, we described how a global potential field map can be dynamically self-organized based on distributed sensory measurements and local proximity associations by a group of autonomous robots. In what follows, we will further discuss how to enable the robots to handle the incremental potential field self-organization task in a more efficient way.

---

[1] *Decision and Control*, John Wiley & Sons Ltd., London, 1966, p 369.

The important feature of our approach is that it optimizes the efficiency of the distance association and map updating by evolving motion strategies for the group robots based on some global fitness criteria. The selected (high-fitness) cooperative motion strategies will be used to control the individual robots in their navigation and sensing. Such an approach to evolution-based cooperative strategy learning and incremental potential map self-organization is referred to as an *evolutionary self-organization approach to collective map building*.

Figure 13.1 illustrates a multi-robot system for evolutionary self-organization. While maintaining and updating a potential field map, a remote host will also be responsible for evolving motion strategies and broadcasting them to the group robots.

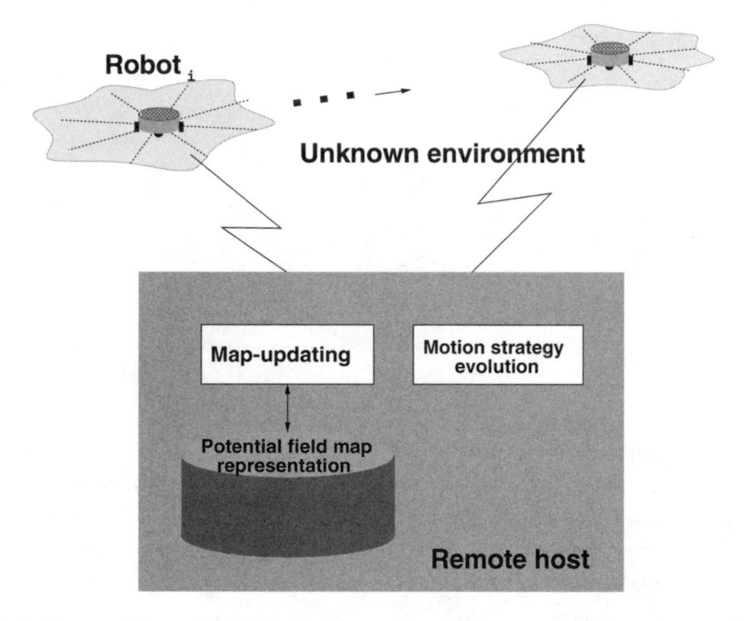

FIGURE 13.1. An illustration of the multi-robot system for evolutionary self-organization.

13.1 Evolution of Cooperative Motion Strategies

Figure 13.2 shows a schematic diagram of the computational architecture for implementing the evolutionary self-organization of a potential field map with a group of autonomous robots. In the process of map building, the motions of the robots are evolved according to two fitness criteria: one is to spatially *diffuse* the robots and another is to *cover* the environment as much as possible, the details of which will be given in Section 13.1.4. The selected motion strategies (stimulus-response pairs) will be performed and evaluated, which in turn serve as the basis for further behavior evolution.

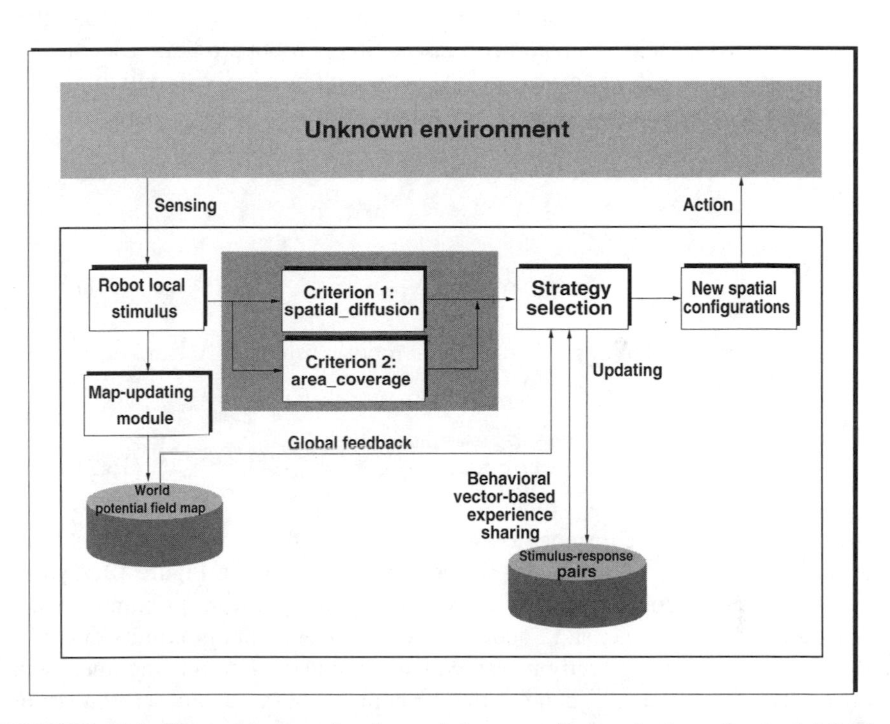

FIGURE 13.2. The architecture for the evolutionary self-organization of potential field maps by a group of autonomous robots.

13.1.1 Representation of a Proximity Stimulus

Generally speaking, the goal of motion strategy evolution is to let group robots learn their reactive motions in response to various sensory stimuli. An acquired stimulus-response pair is expressed in terms of the relative angular difference between the direction of a stimulus and that of a reactive motion. This pair is shared among all the group robots. A detailed definition of the stimulus-response pair will be given in Section 13.1.2.

In our implementation, two situations are considered during the evolution: one is spatial diffusion when the distance between robots i and j, d_{ij}, is less than or equal to a threshold, R_2, and another is area coverage when $d_{ij} > R_2$. In either situation, we use a unified direction representation of robot proximity, denoted by θ_i, to indicate a *significant proximity direction* of all proximity stimuli to robot i. This proximity direction is formally defined as follows:

1. `spatial_diffusion`: For a certain robot i, $\forall m \in [1, M]$ and $m \neq i,\ d_{im} \leq R_2$,

$$\mathtt{vector\_sum}_{m=1}^{m_d-1} \frac{e^{-j\gamma_m}}{d_{im}} \triangleq \zeta_i e^{j\theta_i}, \tag{13.1}$$

2. `area_coverage`: For a certain robot i, $\forall m \in [1, M]$ and $m \neq i,\ d_{im} > R_2$,

$$\mathtt{vector\_sum}_{n=1}^{\mathcal{N}} \frac{e^{j2\pi n/\mathcal{N}}}{D_n^0} \triangleq \zeta_i e^{j\theta_i}, \tag{13.2}$$

where d_{im} denotes the distance between robots i and m. m_d is the number of group robots for spatial diffusion. γ_m denotes the γ angle in Figure 12.3 where robot m occupies location P_j. θ_i corresponds to a significant proximity direction, as mentioned above, and ζ_i corresponds to a significant proximity distance. Both θ_i and ζ_i are not directly measured but computed based on other measured proximity information. D_n^0 denotes the nth component (sector) in sensing vector $\mathcal{I}_0$.

Having identified the two situations in group robots, we can reduce the problem of motion strategy evolution into that of acquiring *two individual reactive motion behaviors*; namely, one for spatial diffusion and another for area coverage, respectively. Both reactive behaviors respond to proximity stimuli as defined in terms of unified significant proximity directions. As opposed to the reactive behavior learning directly for each of possible sensory stimuli, the unified significant proximity representation can improve the efficiency of learning.

13.1.2 Stimulus-Response Pairs

A response evolved in a situation is represented in a vector of varying weights, called a *behavior weight vector*. In other words, with respect to a stimulus represented as a significant proximity direction, θ_i^u, using Eqs. 13.1 and/or 13.2, there exists a probability vector, ϖ_i, where each component expresses the probability of having efficient spatial diffusion and/or area coverage, if a reactive motion in a specific direction is executed. Here the superscript u in θ_i^u indicates that the significant proximity direction θ_i is the uth sector in $[1, \mathcal{N}]$. The pair of θ_i^u and ϖ_i is defined as a *stimulus-response* pair. Specifically, behavior weight vector ϖ_i can be written as follows:

$$\varpi_i \triangleq [\phi_1, \phi_2, \cdots, \phi_k, \cdots, \phi_{\mathcal{N}}], \tag{13.3}$$

where $\phi_k \in [0, 1]$, and $\sum_{k=1}^{\mathcal{N}} \phi_k = 1$.

At each movement step, if robot i has selected the vth direction in $[1, \mathcal{N}]$, denoted by θ_i^v, for its next movement, then the kth component in vector ϖ_i will be updated accordingly as follows:

$$\phi_k^{t+1} = \frac{\phi_k^t + \delta}{1 + \psi}, \tag{13.4}$$

where $\delta = \begin{cases} \psi, & \text{if } k = v - u + 1, \\ 0, & \text{otherwise.} \end{cases}$

ψ denotes a positive coefficient. u and v denote the indices of directions θ_i^u and θ_i^v, respectively, in $[1, \mathcal{N}]$.

Based on a series of updating, the weights for some motion directions become more significant than the others, signifying that motions in the respective directions have higher probabilities of good performance than the rest with respect to a certain stimulus.

13.1.3 Chromosome Representation

At each time step, a robot senses its environment and then derives a significant proximity direction as defined in Eqs. 13.1 and 13.2. Subsequently, it selects a set of possible reactive motion directions based on the experience obtained by its group. That is, a population for evolution is created based on vector ϖ_i. In our present work, once a motion direction is selected, the exact movement step increment within a local region is randomly determined.

In order to represent the next possible movements in our present implementation, we define a two-dimensional Cartesian coordinate frame (x', y') centered at the current location of an individual robot. Next, we consider a square bounding region of width $(2 \cdot d_m + 1)$ around the robot, where d_m denotes the maximum movement step increment for the robot. Thus, the local region for the next movement of the robot is bounded in $x'-$ and $y'-$ directions with respect to the local reference frame, as follows:

$$x', y' \in [-d_m, d_m]. \tag{13.5}$$

Suppose that $(2 \cdot d_m + 1)$ corresponds to a binary string of length L. In such a case, we can uniquely represent a location within the local region with two binary values, each with length L. By the same token, for behavior evolution, we can use a chromosome of length $2L$ to encode the next reactive movement of a robot.

As for the evolution in a group of M robots, the length of a chromosome becomes $(2L) \cdot M$. The chromosome for the group robots is represented in gray codes.

13.1.4 Fitness Functions

In behavior evolution, the fitness function used to select the best next location consists of two terms: one is called *general* fitness and another is called *special* fitness. The general fitness term encourages the group robots to explore the potential field in new, less confident regions, and at the same time avoid repeating the work of other robots. It is defined as follows:

$$s_g = \prod_{i=1}^{m} \left\{ (1 - \mathtt{max}\{w_i^{t_k}\}) \prod_{j=1}^{m_e} \sqrt[4]{d_{ij} - R_1} \right\}, \tag{13.6}$$

where $\mathtt{max}\{w_i^{t_k}\}$ denotes the maximum confidence weight corresponding to the location of robot i. m denotes the number of robots that are grouped together during one evolutionary movement step (of several generations), according to a special fitness term. When the special fitness is concerned with *spatial diffusion*, m becomes m_d. m_e denotes the number of robots that do not belong to m and have just selected and executed their next motion strategies. d_{ij} denotes the distance between robots i and j, which is greater than a predefined distance threshold, R_1.

In addition to the general fitness, we also define two special fitness terms corresponding to the criteria of multi-robot spatial diffusion and area coverage, respectively. They are:

1. `spatial_diffusion`:

$$s_1 = \prod_{i=1}^{m_d - 1} \prod_{j=i+1}^{m_d} \sqrt{d_{ij} - R_2}, \tag{13.7}$$

2. `area_coverage`:

$$s_2 = \frac{\sqrt{\Delta \mathcal{V}}}{\prod_{i=1}^{m_c} \zeta_i}, \tag{13.8}$$

where m_d denotes the number of spatially diffusing robots whose interdistances d_{ij} are greater than distance threshold R_2. $\Delta \mathcal{V}$ denotes the total number of locations visited by a group of m_c area-covering robots based on their selected motion directions. ζ_i denotes a significant proximity distance between robot i and other robots in the environment, as defined in Eq. 13.2.

Having defined the general and the special fitness terms, we can now give the complete fitness function used in the evolution of group robots, as follows:

$$S = \begin{cases} s_g \cdot s_1, & \text{for spatially diffusing robots,} \\ s_g \cdot s_2, & \text{for area-covering robots.} \end{cases} \qquad (13.9)$$

Thus, given a certain stimulus, a robot first of all applies some genetic operations to a population of chromosomes representing possible next locations. This population is created according to Eq. 13.3. Next, the robot selects the fittest member of the population based on its evaluation using Eq. 13.9. Finally, it executes a response motion according to the direction given in the selected chromosome and uses this direction to update the weight vector of Eq. 13.3.

13.1.5 The Algorithm

Figure 13.3 illustrates the scheme of encoding a proximity stimulus with a unified, direction-based representation, evolving a high-fitness next movement, and updating the components of a behavior weight vector based on a selected motion strategy.

The detailed algorithm for the behavior evolution in group robots for the task of collective potential field map building is given in Figure 13.4. At each evolutionary step, a robot selects location P_0^{t+1} based on its current sensory measurement, executes the selected next movement, takes a new measurement, and updates the global potential field map. This cycle repeats itself until the map is accurately constructed.

13.2 Experiments

In order to experimentally validate our multi-robot potential field map building approach, we will examine how a group of autonomous robots in a bounded two-dimensional environment can efficiently perform the task as mentioned in the preceding sections. Figure 13.5 shows the experimental environment and initial spatial distribution of the robots, where symbol $*$ denotes the location of a mobile robot. The environment contains 4 stationary obstacles. All the robots are homed at one corner in the environment. At the beginning, the robots have no *a priori* knowledge about their reactive behaviors.

The parameters used in this experiment are given in Table 13.1. In the table, population size $\mathcal{P}$ and generation size $\mathcal{G}$ are chosen according to the number of group robots, m, involved in the evolution; they are listed with respect to a varying group size of 1 to 6. As the population of reactive motion strategies evolves, the probability of mutation will automatically decrease in a step-by-step fashion.

13.2.1 Experimental Design

For the ease of comparison, we have provided a true potential field map and its contour plot for the given task environment in Figures 13.6(a) and (b),

Parameter	Symbol	Unit	Value
number of robots	M		6
sensory section	$\mathcal{N}$		16
environment size		$grid \times grid$	204×204
map resolution		$grid$	5
map locations			40×40
maximum movement step	d_m	$location$	7
coefficient	η		$\frac{1}{600}$
coefficient	λ		$\frac{1}{5}$
behavior vector increment	ψ		0.2
chromosome length	$(2L) \cdot M$	bit	$16 \cdot M$
population size	$\mathcal{P}$		20/30/45/65/90/120
generations per step	$\mathcal{G}$		8/12/18/26/36/48
crossover probability	p_c		0.6
mutation probability	p_m		0.1/0.05/0.005
distance threshold	R_1	$grid$	10
distance threshold	R_2	$grid$	15

TABLE 13.1. Parameters as used in the experiments.

respectively. When comparing with the true map, we will compute the second-moment errors for all locations in an estimated potential field map at each step of local motion strategy evolution and proximity measurement. The error is formally defined as follows:

$$\varepsilon^t \triangleq \sqrt{\frac{1}{K}\sum_{j=1}^{K}(\mathcal{U}_j^t - \bar{\mathcal{U}}_j)^2}, \tag{13.10}$$

where K denotes the total number of locations in the potential field map, and $\mathcal{U}_j^t$ and $\bar{\mathcal{U}}_j$ denote the estimated and the true potential values at location $P_j(x_j, y_j)$, respectively.

13.2.2 Comparison with a Non-Evolutionary Mode

In order to evaluate the performance of the evolutionary approach to incremental potential field map building, we also compare a resulting map with the one generated by a group of autonomous robots that utilized predefined reactive motion strategies without any behavior evolution (a non-evolutionary mode). In both evolutionary and non-evolutionary modes, we use the same set of parameters as given in Table 13.1.

In the non-evolutionary mode, individual robots select and execute their local motions by calculating the standard deviation of potential field values within their neighboring regions of radius d_m (the maximum movement step increment). The next movement at time $t + 1$ will be determined by motion direction ξ and step increment d_s. Thus, the location at time $t + 1$, P_0^{t+1}, can be described as follows:

$$P_0^{t+1} = P_0^t + d_s \cdot e^{j\xi}. \tag{13.11}$$

More specifically, a vector Δ is defined to record the standard deviation of potential value differences between times t and $t - 1$, for all *sensing sectors*. The vth component of this vector is calculated as follows:

$$\Delta_v = \mathtt{std}(\{\iota_{vj} \mid \iota_{vj} = \mathcal{U}_{vj}^t - \mathcal{U}_{vj}^{t-1},\ \forall j \in \mathcal{E}_v\}),\ v = 1, 2, \cdots, \mathcal{N}\}), \tag{13.12}$$

where operator `std` denotes the standard deviation for all locations inside the vth sensing sector, $\mathcal{E}_v$.

In addition, another vector Λ is also defined to store the standard deviation of potential values at time t around each *location* within the same sensing sector. The jth component of this vector is determined as follows:

$$\Lambda_j = \mathtt{std}(\{v_{jl} \mid v_{jl} = \mathcal{U}_{jl}^t,\ \forall l \in \mathcal{E}_v,\ |l - j| \leq 1\}), \tag{13.13}$$

where $\mathcal{E}_v$ is determined based on Eq. 13.12.

Based on the above definitions, robot i in a non-evolutionary mode chooses its next movement direction θ_i^v (the vth sensing sector) whenever the following is satisfied:

$$\theta_i^v|_{\Delta_v = \max(\Delta_1, \Delta_2, \cdots, \Delta_{\mathcal{N}})}, \tag{13.14}$$

and

$$\forall k,\ P_k \notin \mathcal{E}_v, \tag{13.15}$$

where operator `max` returns the maximum from a set of values. P_k denotes the position of robot k.

Having determined its movement direction sector, the robot further chooses its next location $P_0^{t+1}(x_0, y_0)$ to move to, within the chosen sector. This location should satisfy the following condition:

$$(x_0, y_0)|_{\Lambda_j(x_0, y_0) = \max(\Lambda_1, \Lambda_2, \cdots)}. \tag{13.16}$$

13.2.3 Experimental Results

Figures 13.7 and 13.8 present the potential field maps obtained in the evolutionary and non-evolutionary modes, respectively, along with their contour plots. The robot motion trajectories in the two modes are shown in Figures 13.9 and 13.10, respectively. In the evolutionary mode of Figure 13.9, the robots collectively build a potential field map in about 10 movement steps. Their navigation tends to explore uncovered areas first and then try to refine local potential field values. On the other hand, in the non-evolutionary mode of Figure 13.10, the robots are more cluttered together during their motions. This wastes some searching time for the group. Hence, it takes a longer time for the non-evolutionary robots to make proximity measurements and associations. If we compare these two sets of results, we can readily note that the group of evolutionary robots leads to a faster convergence of the global potential field map.

In order to examine the overall spatial distributions of group robots during map building, Figures 13.11 and 13.12 present the locations in the unknown environment visited by the robots in two different modes, respectively. It is interesting to observe that the group robots with behavior evolution can sample the unknown environment evenly, along the valley of the potential field. They focus slightly more on the junction locations in order to eliminate the proximity uncertainty involved. In the non-evolutionary mode, however, the robots quite often go to the locations near the edges of the obstacles where large standard deviations of the map are found.

In addition to the above observations, Figure 13.13 further provides a second-moment error comparison between the two modes of map building. It is obvious from the figure that the error in the case of evolutionary robots goes down much faster than the one in the non-evolutionary case.

13.3 Discussions

In what follows, we will make several remarks about the evolutionary self-organization of a potential field map by a group of autonomous robots.

13.3.1 Evolution of Group Behaviors

Figure 13.14 presents the initial motion trajectories of 6 robots having no *a priori* knowledge about their reactive motion strategies. The robots start with a phase of exploratory world modeling, in which the robots interact with their environment by executing some S-shaped movements. As a result, robots 4, 6, and 1 move to regions B, C, and D, respectively.

The process of evolving group behaviors during potential field map building can be readily observed from Figure 13.15 (where $t_n^{(m)}$ signifies robot *m* at step n). We note that the trajectories of some robots become smoother, as some stable behavior responses are acquired by the robots. At the same time, we also note from Figure 13.16 that the remaining two robots (robots 3 and 5) concentrate more on the finer details in the areas omitted by the other 4 robots.

13.3.2 Cooperation among Robots

In Figure 13.17, some degree of cooperation between robots 1 and 6 can be observed as the two are moving closer to each other at step 6. As far as robot 1 is concerned, it turns around to visit other uncovered locations at step 7, leaving the locations ahead to robot 6. However, at step 8, it returns back to the locations on the right, as robot 6 moves away. A similar behavior is also found in robot 6. Such a cooperative group motion strategy has been found quite effective in exploring the unknown environment and building a global potential map along the way.

Figure 13.18 provides a zoom-in view of Figure 13.15, from which we notice that at step 4, robot 2 encounters two choices of motion: one is to go from region B to F and another is to E. However, since region F has already been visited by robot 4, robot 2 decides to move toward region E. So does robot 4 at step 5, moving from region G toward H.

13.4 Summary

In this chapter, we described an evolutionary computation approach to the emergence of group behaviors for incrementally self-organizing a global potential field representation in an unknown environment. While giving the underlying modeling and computation formalisms, we have presented several results from our experimental validation. Generally speaking, our approach enables the distributed robots to gradually develop an ability of experience-based cooperation and adaptation to their task environment.

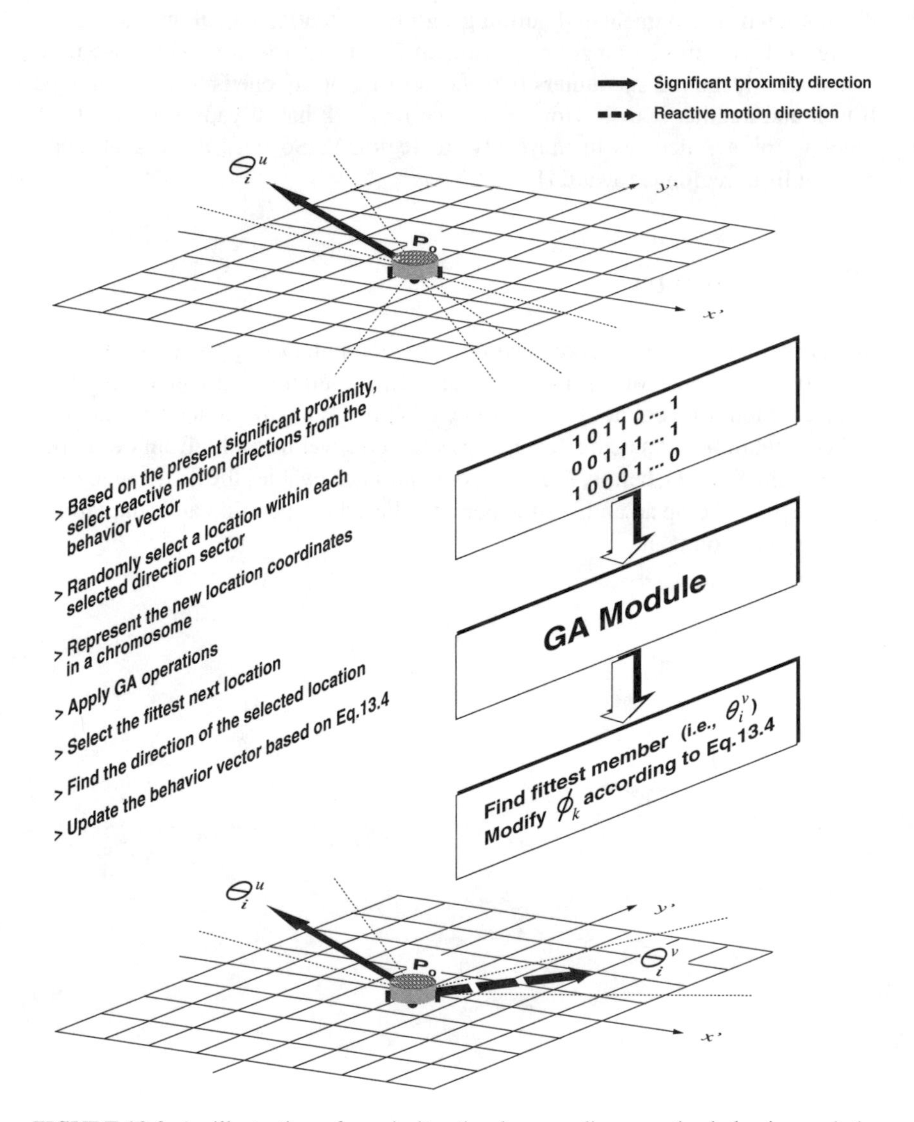

FIGURE 13.3. An illustration of proximity stimulus encoding, reactive behavior evolution, and behavior vector updating.

```
begin
define fitness function S (i.e., general and special terms) for the current step,
define the maximum number of generations per step G w.r.t. robot group size,
define population size P w.r.t. the robot group size,
define crossover probability p_c,
define mutation probability p_m,
for generation : 1 ⟶ G do
   for robots of the same group (e.g., spatial diffusion or area coverage)
      sense and compute stimulus directions,
      find possible reactive directions based on stimulus-response pairs,
      randomly select locations in their respective directions,
      form a chromosome by combining their motion strategies,
      create population P of possible group motion strategies,
   endfor
   use two-point crossover with probability p_c,
   mutate the members of generation with probability p_m,
   evaluate the current generation according to fitness function S:
      for population: 1 ⟶ P do
         modify robot positions w.r.t. the chromosome of population,
         compute the fitness function S for the group robots,
      endfor
   select the best reactive motion strategies based on computed fitness values,
   execute the selected reactive motions by the group robots,
   update behavior vector ϖ_i, taking into account the executed motions,
endfor
end
```

FIGURE 13.4. The algorithm for evolving reactive motion strategies for group robots with respect to their stimuli.

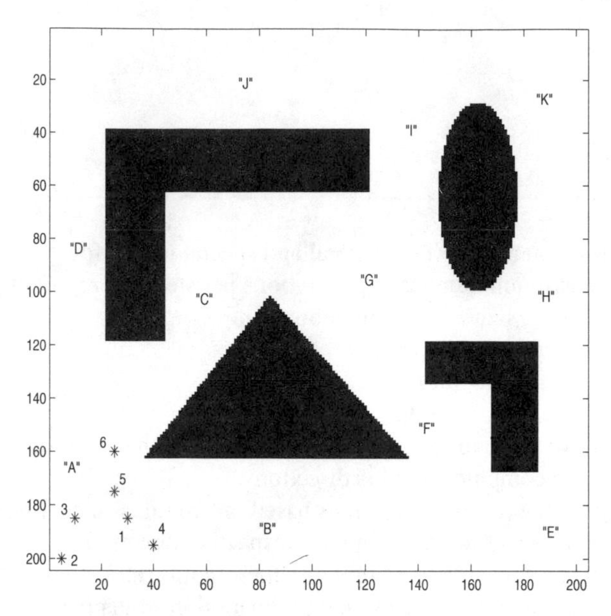

FIGURE 13.5. The environment used for experimentation. The numbers signify the initial locations of group robots, and the capital letters are regional labels (©1999 IEEE).

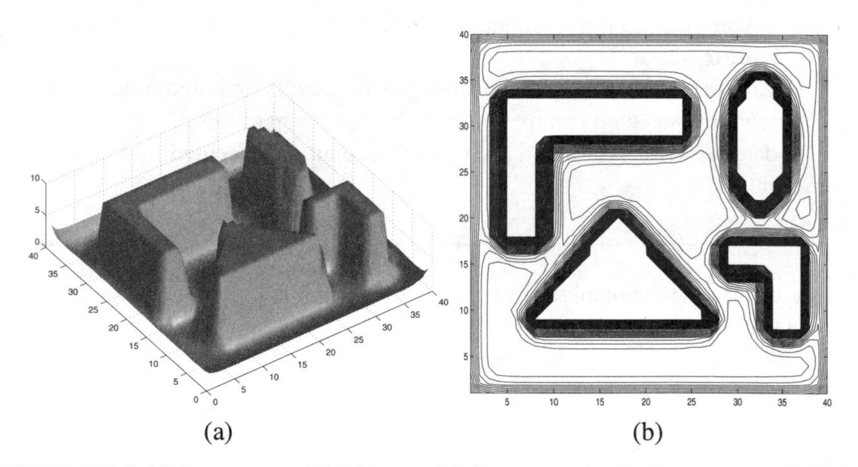

FIGURE 13.6. (a) A true potential field map. (b) A contour plot for the true potential field map (©1999 IEEE).

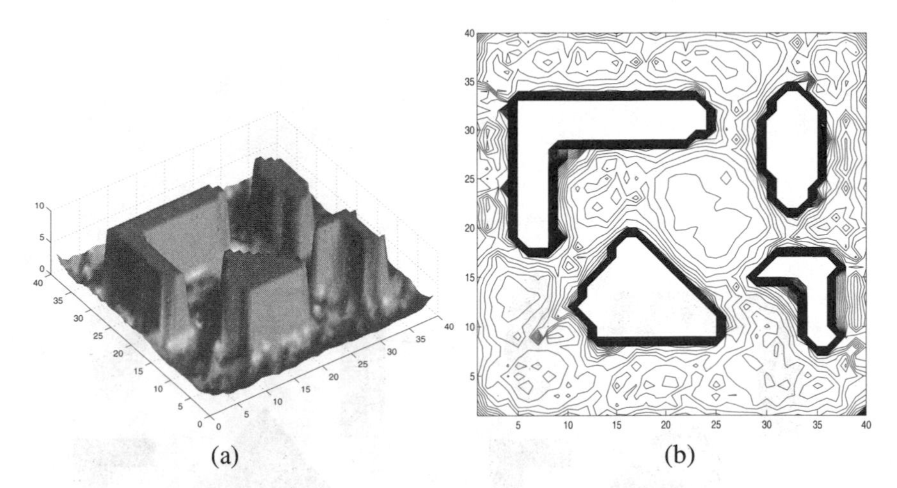

FIGURE 13.7. *Evolutionary mode:* (a) A potential field map built by 6 group robots after 20 movement steps. (b) A contour plot for the obtained potential field map (©1999 IEEE).

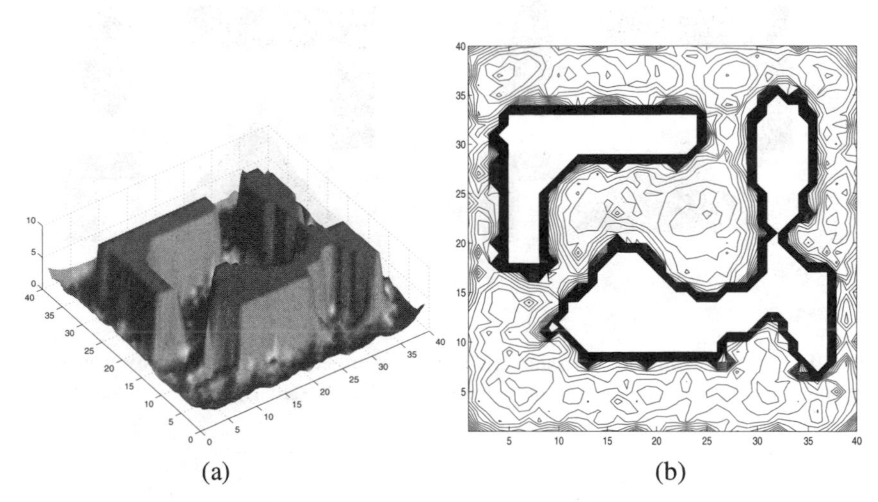

FIGURE 13.8. *Non-evolutionary mode:* (a) A potential field map built by 6 robots with predefined reactive motion strategies after 24 movement steps. (b) A contour plot for the obtained potential field map (©1999 IEEE).

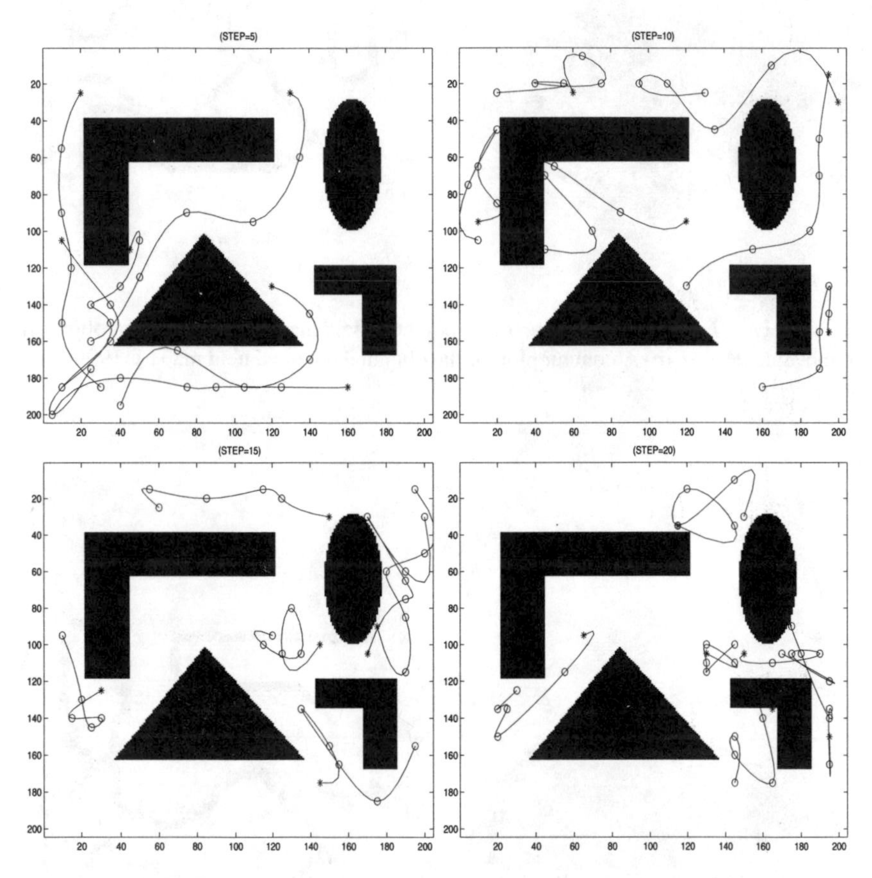

FIGURE 13.9. *Evolutionary mode:* The motion trajectories produced by a group of evolutionary robots in the first 20 movement steps (©1999 IEEE).

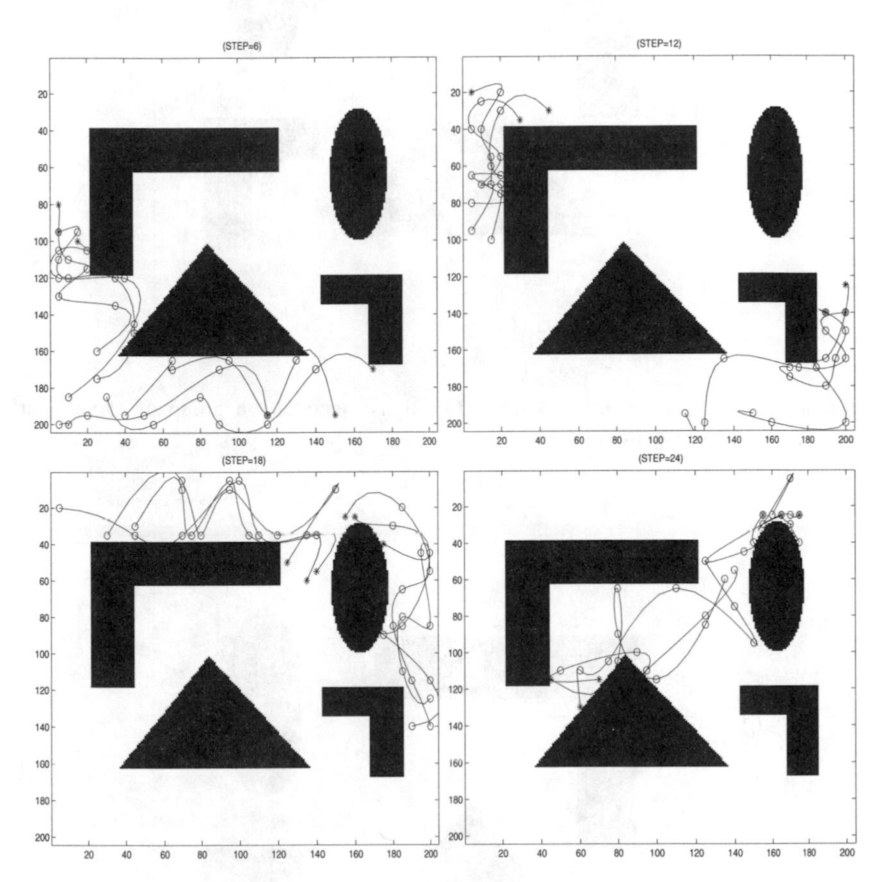

FIGURE 13.10. *Non-evolutionary mode:* The motion trajectories produced by a group of non-evolutionary robots in the first 24 movement steps (©1999 IEEE).

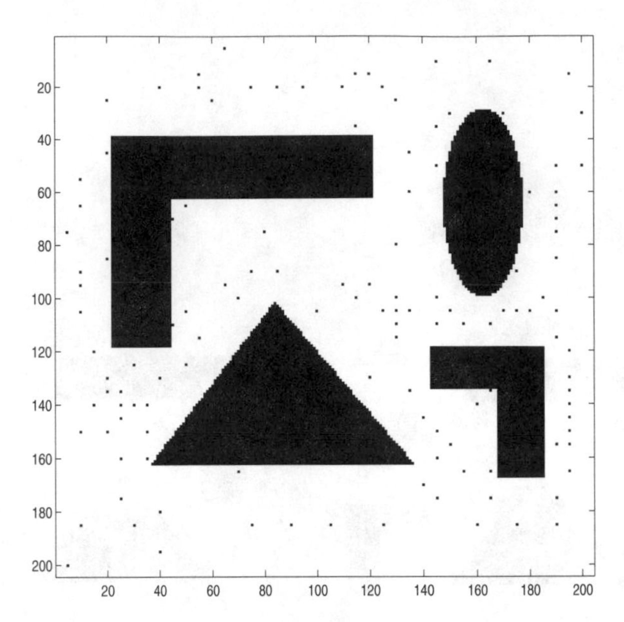

FIGURE 13.11. *Evolutionary mode:* Locations visited by a group of evolutionary autonomous robots (©1999 IEEE).

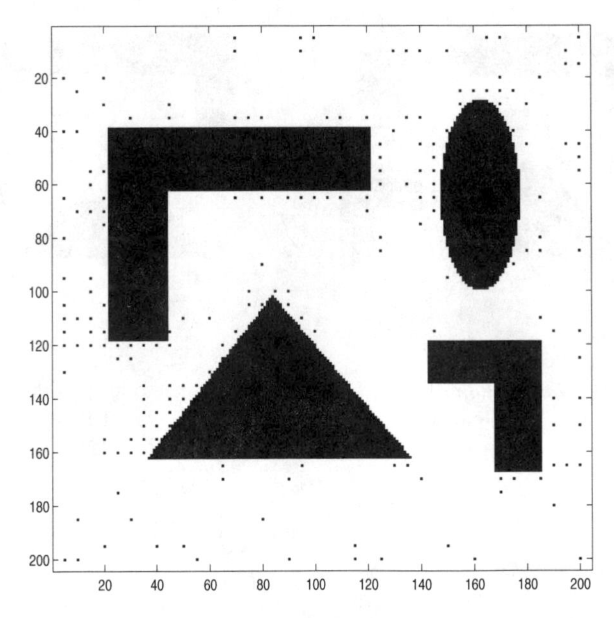

FIGURE 13.12. *Non-evolutionary mode:* Locations visited by a group of non-evolutionary robots (©1999 IEEE).

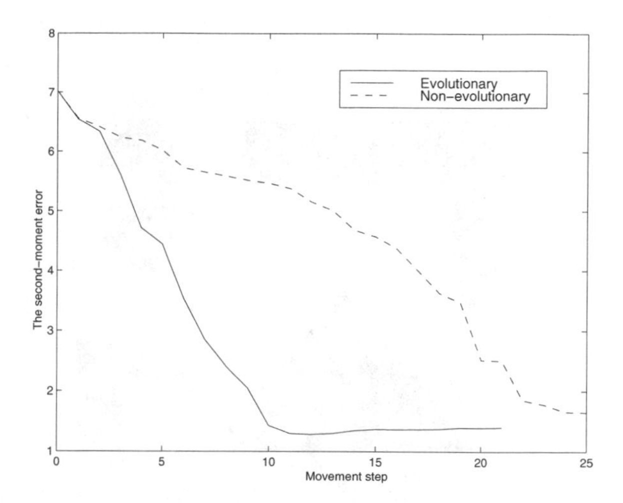

FIGURE 13.13. A second-moment error comparison between the two modes of map building. The solid line corresponds to the error measured in the case of evolutionary robots during their evolutionary self-organization of a potential field map, whereas the dashed line corresponds to that in the case of non-evolutionary robots (©1999 IEEE).

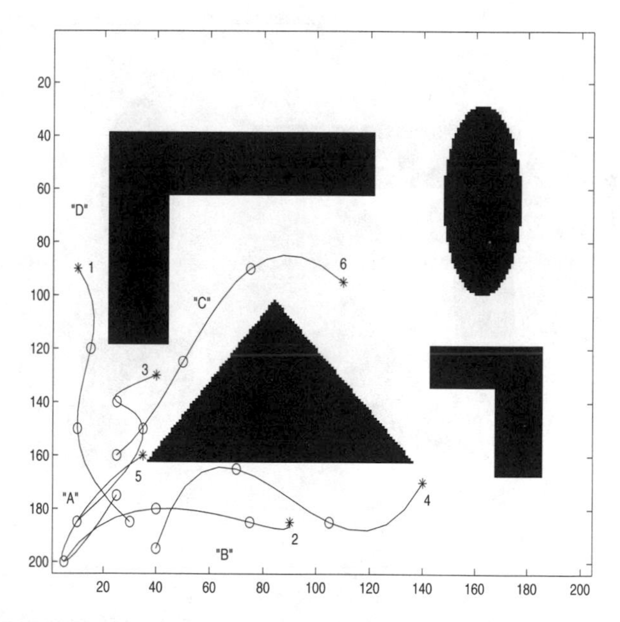

FIGURE 13.14. The initial motion trajectories of 6 robots in an unknown environment.

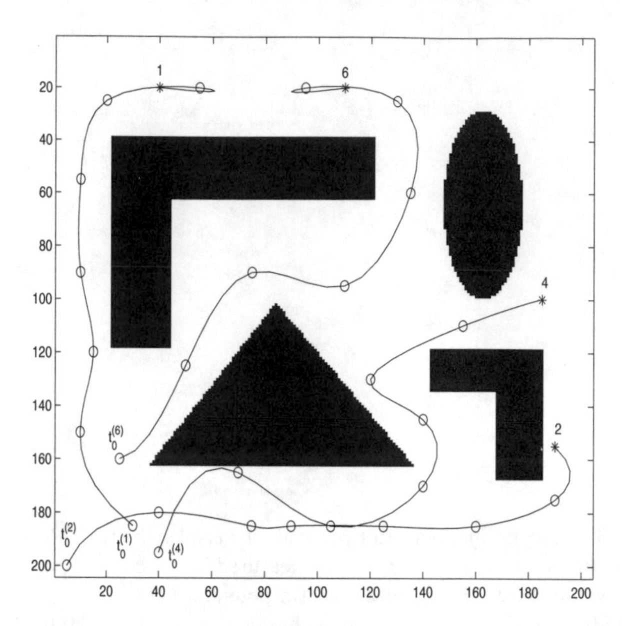

FIGURE 13.15. The motion trajectories of 4 robots in an unknown environment.

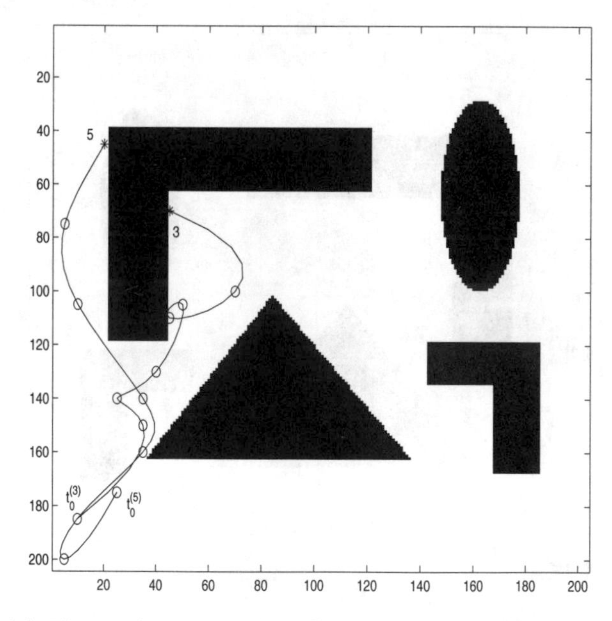

FIGURE 13.16. The motion trajectories of 2 robots focusing on the details of the environment.

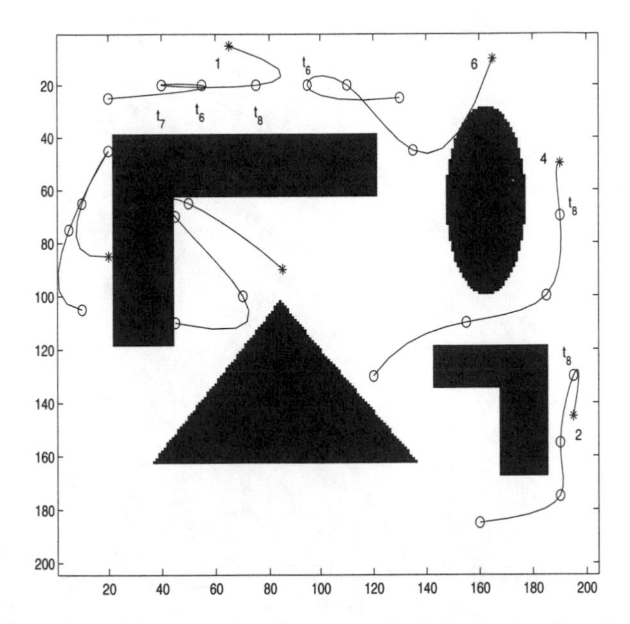

FIGURE 13.17. The emergence of a cooperative behavior in robots 1 and 6.

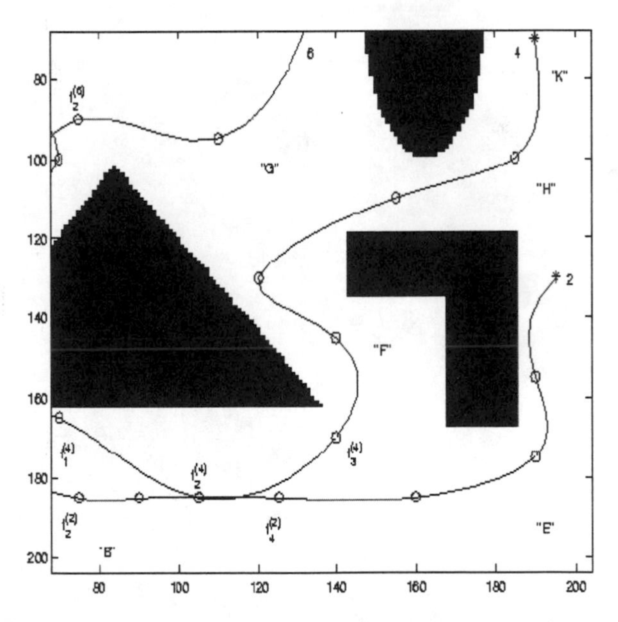

FIGURE 13.18. The emergence of a cooperative behavior in robots 2 and 4.

Part V

An Exploration Tool

14
Toolboxes for Multi-Agent Robotics

Nor can we know or imagine now the limitless beauty yet to be revealed in the future – by science.[1]

Isaac Asimov

14.1 Overview

MATLAB[2] is a computing environment for high-performance numerical computation and visualization. MATLAB integrates numerical analysis, matrix computation, signal processing, and graphics in an easy-to-use environment. MATLAB

[1] *The Roving Mind*, Prometheus Books, 1983.
[2] MATLAB is a trademark of MathWorks, Inc.

also features a family of application-specific solutions that are called *toolboxes*. They are comprehensive collections of MATLAB functions (M-files) that extend the MATLAB environment in order to solve particular classes of problems. Areas in which toolboxes are available include signal processing, control systems design, dynamic system simulation, systems identification, image processing, neural networks, and others.

In this chapter we present a set of new toolboxes, called *Multi-Agent Robotics Toolboxes*,[3] that we developed for conducting research studies related to multi-robot reinforcement learning, adaptation, self-organization, trajectory generation, and graphics. The toolboxes include:

- Multi-Agent Reinforcement Learning
- Evolutionary Multi-Agent Reinforcement Learning
- Evolutionary Collective Behavior Implementation
- Multi-Agent Self-Organization
- Evolutionary Multi-Agent Self-Organization

These toolboxes are useful for simulation as well as for analyzing results of experiments with simulated robots. They are based on a very general method of representing the evolutionary approaches to multi-agent robotic systems by *description matrices*.

In what follows, we provide detailed references on the architecture, file structure, function description, user configuration, and data structure used in each of the toolboxes.

14.2 Toolbox for Multi-Agent Reinforcement Learning

14.2.1 Architecture

The architecture for implementing multi-agent reinforcement learning is given in Figure 14.1.

14.2.2 File Structure

The file structure for the multi-agent reinforcement learning toolbox is shown in Figure 14.2. The calling tree for the system initialization in this toolbox is shown in Figure 14.3, where the functions in shaded blocks are user defined.

[3] http : //www.crcpress.com/us/ElectronicProducts/downandup.asp?mscssid =

14.2.3 Function Description

The toolbox functions for multi-agent reinforcement learning and for data processing are described in Tables 14.1 and 14.2, respectively.

14.2.4 User Configuration

set_default_value:

1. `Current_X` and `Current_Y` define the central point for display. `Visible_Width` defines the size of the area for graphics display.

2. `Region_Max` sets the width of a square area for the experiments. Together with other four parameters, `Xmin`, `Ymin`, `Xmax`, and `Ymax`, an experimental area can be accurately defined.

3. `Origin_Sense_Radius` defines the initial range of a sensor.

4. `Sensing_Step_Ratio` determines the step size for each robot. It defines the relationship between a sensing radius and a step size, i.e.,

$$d_s = \texttt{Sensing\_Step\_Ratio} \cdot R_0, \tag{14.1}$$

   where R_0 is the sensing radius of a robot and d_s is the step size.

5. `Delta_Radius` models a robot sensor as an incremental device. Before it can find its operating object, a robot increases its sensing range gradually with the increment of `Delta_Radius`.

6. `Move_Step` defines the movement step of the objects in the experiments.

7. `Cell_Min_Distance` indicates the minimum distance between robots.

8. `Red_White_Distance` defines the minimum safety distance between a robot and an object.

9. `Break_Point` sets the maximum number of program running time steps.

10. `Increment_Now`, `Distribute_Now`, `Selection_Now`, `PostProcess_Now`, and `Region_Num` are all initial experimental settings to configure the weight increment, distribution mode, behavior selection strategy, post-processing style, and resolution scale, respectively.

main_ini:

The initialization for this file is similar to that for the same file in the toolbox for *Evolutionary Collective Behavior Implementation*, as will be described later. So is the initialization for `set_norm_class.m` and `char_position_def.m`.

14.2.5 Data Structure

In this toolbox, matrices `CLASS`, `CELL`, and `HISTORY` have the same forms as the ones given in Figures 14.9, 14.10, and 14.11, respectively. The representations of `Red_Step_Matrix`, `Red_Position`, and `LoopEnd_Behavior` have the same format as shown in Figure 14.4.

Assume that there are K classes of robots involved in the experiments, and the number of each class of robots are M_1, M_2, $\cdots$, M_K, respectively. Here, $M_{cel} = \sum_{i=1}^{K} M_i$.

1. `Red_Step_Matrix`: For this matrix, $M = M_{cel}$. The submatrices in Figure 14.4 can be expressed as follows:

$$[d_{11} \ \cdots \ d_{ij} \ \cdots \ d_{KN_K}], \tag{14.2}$$

   where d_{ij} is the step of the cell (robot) j of class i at time t.

2. `Red_Position`: Here, $M = 2 \cdot M_{cel}$. The submatrices in Figure 14.4 can be expressed as follows:

$$[x_{11} \ y_{11} \ \cdots \ x_{ij} \ y_{ij} \ \cdots \ x_{KN_K} \ y_{KN_K}], \tag{14.3}$$

   where $(x_{ij}, \ y_{ij})$ is the position of the cell (robot) j of class i at time t.

3. `LoopEnd_Behavior`: This matrix records the history of weights corresponding to the primitive behavior set. Assume that there are $\mathcal{B}$ primitive behaviors for each robot. Then for `LoopEnd_Behavior`, $M = \mathcal{B}$. The submatrix in Figure 14.4 can be expressed as follows:

$$[w_1 \ w_2 \ \cdots \ w_{\mathcal{B}}], \tag{14.4}$$

   where $\sum_{i=1}^{\mathcal{B}} w_i = 1$.

14.3 Toolbox for Evolutionary Multi-Agent Reinforcement Learning

14.3.1 File Structure

The file structure for the evolutionary multi-agent reinforcement learning toolbox is shown in Figure 14.5. The calling tree of this toolbox is given in Figure 14.6, where the functions in shaded blocks are user defined.

14.3.2 Function Description

The toolbox functions for evolutionary multi-agent reinforcement learning are described in Table 14.3.

14.3.3 User Configuration

initial_para:

1. `Fitness_Fun` corresponds to the name of the fitness function.
2. `Range_Chrom` defines the range of chromosomes. In some cases, the domain of definition for genetic operations is a subset of all possible chromosomes.
3. `Bit_Length` sets the bit length of the chromosomes in genetic optimization.
4. `Population_Size` defines the size of the population in genetic operations.
5. `Total_Generations` sets the number of generations in genetic optimization.
6. `Cross_Prob` and `Mutation_Prob` define the probabilities of crossover and mutation in genetic operations, respectively.
7. `Analysis_Time` sets the steps for analysis in the experiments.
8. `Case_Default` and `Case_Position` initialize the settings associated with the results of *multi-agent reinforcement learning* experiments.

14.4 Toolboxes for Evolutionary Collective Behavior Implementation

14.4.1 Toolbox for Collective Box-Pushing by Artificial Repulsive Forces

14.4.1.1 File Structure

The file structure for the toolbox of collective box-pushing by artificial repulsive forces is shown in Figure 14.7. The calling tree of this toolbox is presented in Figure 14.8, where the functions in shaded blocks are user defined.

14.4.1.2 Function Description

The toolbox functions for collective behavior based on artificial repulsive forces and for data processing are described in Tables 14.4 and 14.5, respectively.

14.4.1.3 User Configuration

char_position_def:

In this file, the positions of all features to model a class in matrix `CLASS` are defined. They act as an index for later display, computation, etc. The corresponding values of the features will be assigned in `set_norm_class.m`.

ga_initial:

1. RadiusCode_Len and AngleCode_Len specify the bit lengths of binary representations for a moving radius and an angle in genetic chromosomes, respectively.

2. Fitness_Fun corresponds to the name of a fitness function.

3. Population_Size defines the size of a population in genetic operations.

4. Total_Generations indicates the number of generations in evolution for each movement of robots.

5. Cross_Prob and Mutation_Prob define the probabilities of crossover and mutation in genetic operations, respectively.

6. BreakGA_Ratio is used to break the genetic optimization when the specified proportion of individuals in the whole population has converged.

main_ini:

1. Behav_Matrix_Size defines the size of the behavior matrix.

2. Total_Goal_Num sets the number of goals in the experiment.

3. Class_Num indicates the number of classes. All the classes are numbered in Class_Id_Vector one by one.

4. Char_Matrix_Size defines the size of a matrix that characterizes the features of the classes.

replsmain:

1. FLname assigns a file name and a destination/path for saving the experimental results.

set_default_value:

1. Total_Cell_Num defines the number of robots in the experiment.

2. Cell_Name numbers the robots in a vector of size $1 \times$ Total_Cell_Num.

3. Robot_Distrib defines a square area for the initial distribution of robots. For example, when Robot_Distrib $= [a\ b]$, the initial positions of robots are in the area of a square with 4 corners at $(a,\ a)$, $(a,\ b)$, $(b,\ b)$, and $(b,\ a)$. Object_Distrib and Goal_Distrib define 2 square areas for the initial distributions of a box and a goal, respectively, with the same strategy as Robot_Distrib.

4. Goal_Vary_Tm defines the interval to change the location of the goal.

5. `Step_Ratio` is the ratio of a step size to a repulsive force, i.e.,

$$d_s = \texttt{Step\_Ratio} \cdot F, \tag{14.5}$$

   where d_s is the step size and F is the repulsive force.

6. `Step_Change_Ratio` is an ideal ratio for robot evolution. A box is expected to be pushed at the step that is inverse to the distance between the box and a goal.

7. `Cell_Min_Distance` indicates the minimum distance between robots.

8. `Red_White_Distance` defines the limited distance between a robot and a box.

9. `Class_Change_Dist` sets the distance for robot-box interaction.

10. `Region_Num` specifies the number of regions divided equally around robots.

11. `Break_Point` sets the maximum number of program running time steps.

set_norm_class:

1. `Character` needs to be initialized one by one for each class. The meaning of each element is defined in `char_position_def.m`.

14.4.1.4 Data Structure

Assume that there are K classes of robots involved in the experiments, and the numbers for different classes of robots are M_1, M_2, $\cdots$, M_K, respectively.

1. `CLASS`

   `CLASS` is a matrix that indicates the features and characteristic parameters for all classes. It has the structure given in Figure 14.9.

   $\texttt{C}_i$ is a submatrix for class i. In this submatrix, part I records the class name (identification); part II, $\texttt{B}_i$, defines the behavior parameters; part III, $\texttt{G}_i$, is for goal definition; and all the other feature expressions of this class are in part IV, $\texttt{A}_i$. The total row number of $\texttt{C}_i$ is equal to the maximum of the numbers of rows in $\texttt{B}_i$, $\texttt{G}_i$, and $\texttt{A}_i$. Note that some of the elements in the matrix may not be used. In such a case, the unused elements are assigned with -1, as illustrated in Figure 14.9. Part I is a one-column vector, as follows:

$$\begin{bmatrix} i \\ i \\ \vdots \\ i \end{bmatrix}. \tag{14.6}$$

2. CELL

   CELL defines the positions of robots of all classes in an environment. It has the form given in Figure 14.10, where $M_{cel} = \sum_{i=1}^{K} M_i$, and for the jth cell (robot) of class i, the vector in CELL is:

$$[i \ j \ x_{ij} \ y_{ij}], \tag{14.7}$$

   where $(x_{ij}, \ y_{ij})$ is the coordinate of this robot in the experimental environment.

3. HISTORY

   The movements of all robots are recorded in HISTORY, as shown in Figure 14.11. At time t, the submatrix of HISTORY, $\mathtt{H}_i$, contains the information about the current positions of all robots. $\mathtt{H}_i$ organizes them one by one corresponding to CELL, i.e., $\mathtt{H}_i$ has M_{cel} too. The historical position of the jth cell (robot) of class i at time t is given by:

$$[i \ j \ t \ x_{ij}^t \ y_{ij}^t], \tag{14.8}$$

   where $(x_{ij}^t, \ y_{ij}^t)$ is the coordinate of this robot in the experimental environment at time t.

14.4.2 Toolbox for Implementing Cylindrical/Cubic Box-Pushing Tasks

14.4.2.1 File Structure

The file structure for the toolbox that implements collective cylindrical box-pushing and cubic box-pushing is presented in Figure 14.12. The calling tree of this toolbox is shown in Figure 14.13, where the functions in shaded blocks are user defined.

14.4.2.2 Function Description

The toolbox functions for evolutionary cylindrical/cubic box-pushing and for data processing are described in Tables 14.6 and 14.7, respectively.

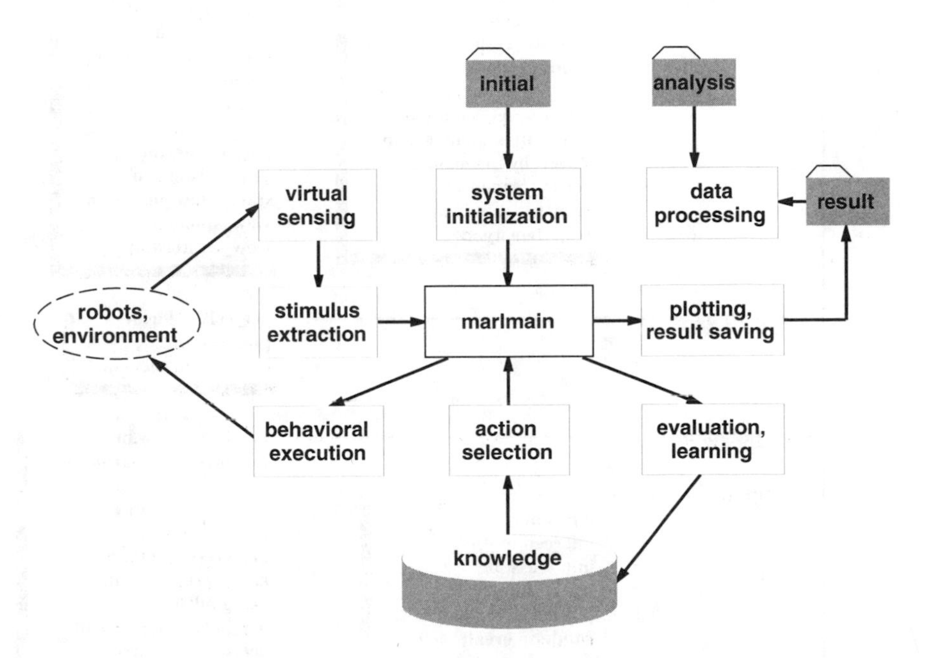

FIGURE 14.1. The architecture of the multi-agent reinforcement learning toolbox.

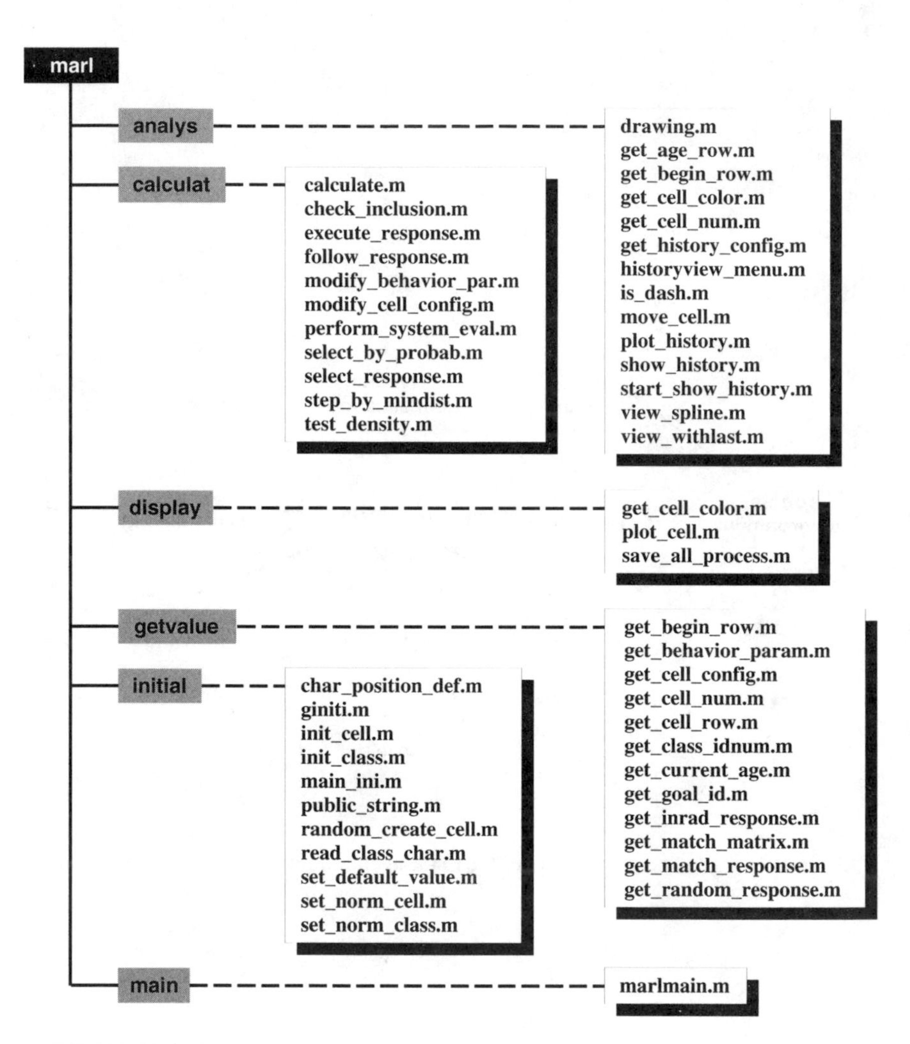

FIGURE 14.2. The file structure for the multi-agent reinforcement learning toolbox.

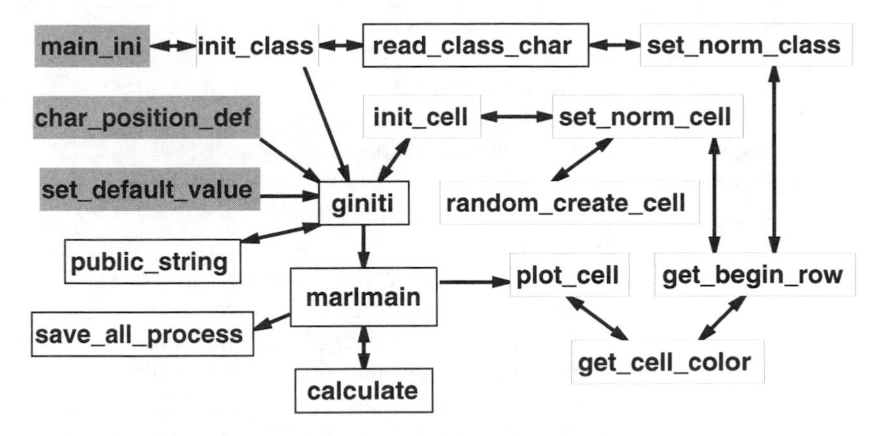

FIGURE 14.3. The calling tree for the initialization in the multi-agent reinforcement learning toolbox.

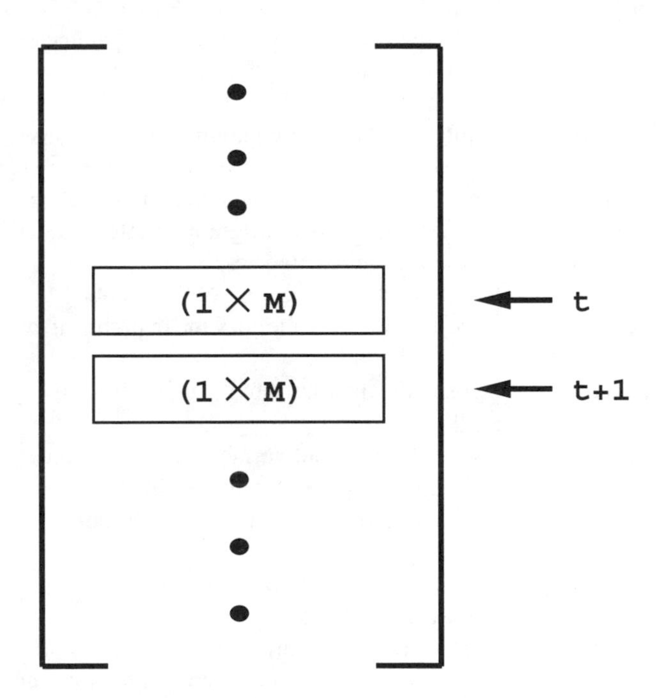

FIGURE 14.4. A schematic diagram illustrating matrices `Red_Step_Matrix`, `Red_Position`, and `LoopEnd_Behavior`.

Function	Description
calculate	perform the main computation
char_position_def	define the position of feature parameters
check_inclusion	determine which region the cell (robot) is in
execute_response	execute the selected response
follow_response	put the selected response to use
get_begin_row	obtain the beginning row of a class in CLASS
get_behavior_param	obtain submatrix with behavior parameters from CLASS
get_cell_color	obtain the display color of a robot from CLASS
get_cell_config	obtain the configuration of the current robot
get_cell_num	obtain the robot number of the current class
get_cell_row	obtain the offset of the current robot in CELL
get_class_idnum	obtain the number of a certain class
get_current_age	obtain the current time step of the robot from HISTORY
get_goal_id	obtain the goal identification in CLASS
get_inrad_response	obtain a behavior identification
get_match_matrix	create a matrix including class/goal name and a stimulus
get_match_response	obtain the response that matches the current stimulus
get_random_response	select a response from CLASS randomly
giniti	initialize overall experimentation parameters
init_cell	initialize CELL
init_class	initialize CLASS
main_ini	initialize the system parameters
marlmain	serve as the main program for reinforcement learning
modify_behavior_par	modify the behavior weight and normalize it
modify_cell_config	modify the robot configuration after evaluation
perform_system_eval	evaluate the selected response
plot_cell	plot the current location of the robots
public_string	create a string to identify the experiment and the file name for saving
random_create_cell	create the distribution of robots randomly
read_class_char	fill in CLASS
save_all_process	save all important variables into MAT-file
select_by_probab	choose a response by probability
select_response	obtain a stimulus for the current robot and select a sub-behavior
set_default_value	set some variables with default value
set_norm_cell	initialize CELL for classes
set_norm_class	set the content of an environment for CLASS
step_by_mindist	calculate the minimum step in a forward movement
test_density	detect the distribution of robots

TABLE 14.1. Description of the functions in the toolbox for implementing multi-agent reinforcement learning.

Function	Description
drawing	draw the experimental results
get_age_row	obtain the row of age in CLASS
get_begin_row	obtain the beginning row of a class
get_cell_color	obtain the cell (robot) color from CLASS
get_cell_num	obtain the robot number of the current class
get_history_config	match the configuration of the robot in history
historyview_menu	control the menu in the history window
is_dash	test whether there is a dash symbol in a file name
move_cell	move the robot in a color map
plot_history	draw the robot in HISTORY
show_history	show the process in HISTORY
start_show_history	start to show history
view_spline	plot the spline trajectories of robots
view_onetime	observe the experimental situation at a time step

TABLE 14.2. Description of the functions for analyzing the results of multi-agent reinforcement learning.

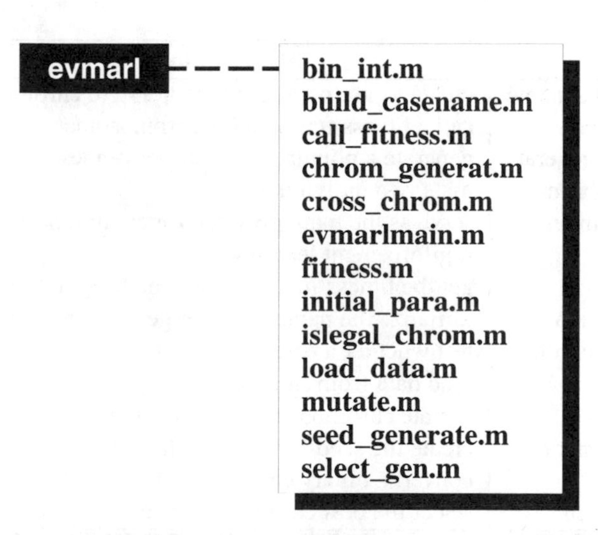

FIGURE 14.5. The file structure for the evolutionary multi-agent reinforcement learning toolbox.

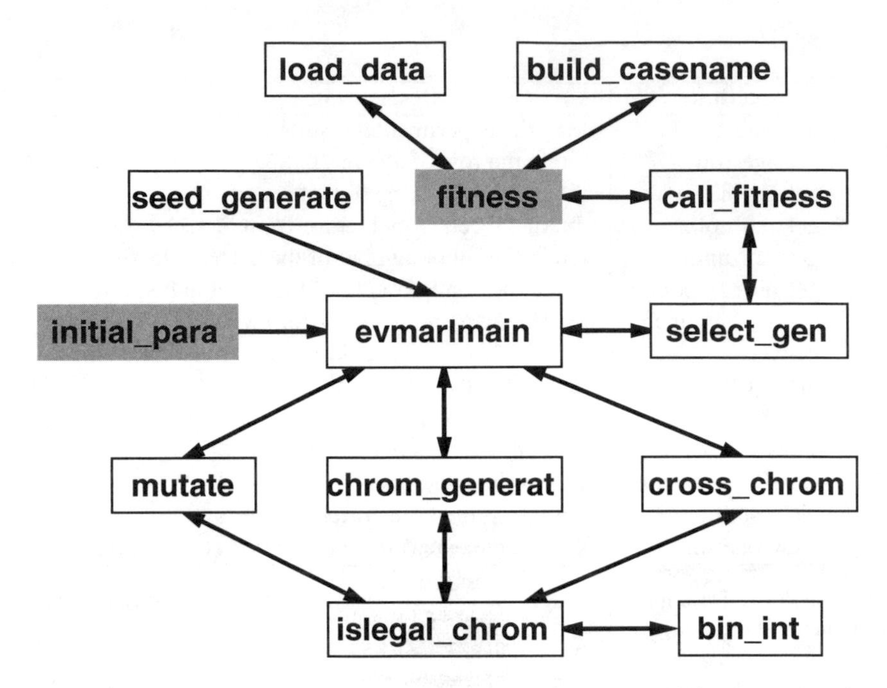

FIGURE 14.6. The calling tree for the evolutionary multi-agent reinforcement learning toolbox.

Function	Description
build_casename	create a case name corresponding to chromosomes
call_fitness	call a fitness function for chromosomes
chrom_generat	generate a population of chromosomes
cross_chrom	mate two individuals
evmarlmain	serve as the main program for evolutionary reinforcement learning
fitness	get the fitness function for genetic operations
initial_para	initialize the parameters for genetic operations
islegal_chrom	test whether a chromosome is legal
load_data	load data from storage
mutate	mutate chromosomes at a given rate
seed_generate	create the seed for random data generation
bin_int	convert a binary code to an integer number
select_gen	select the best chromosomes for the next generation

TABLE 14.3. Description of the functions in the evolutionary multi-agent reinforcement learning toolbox.

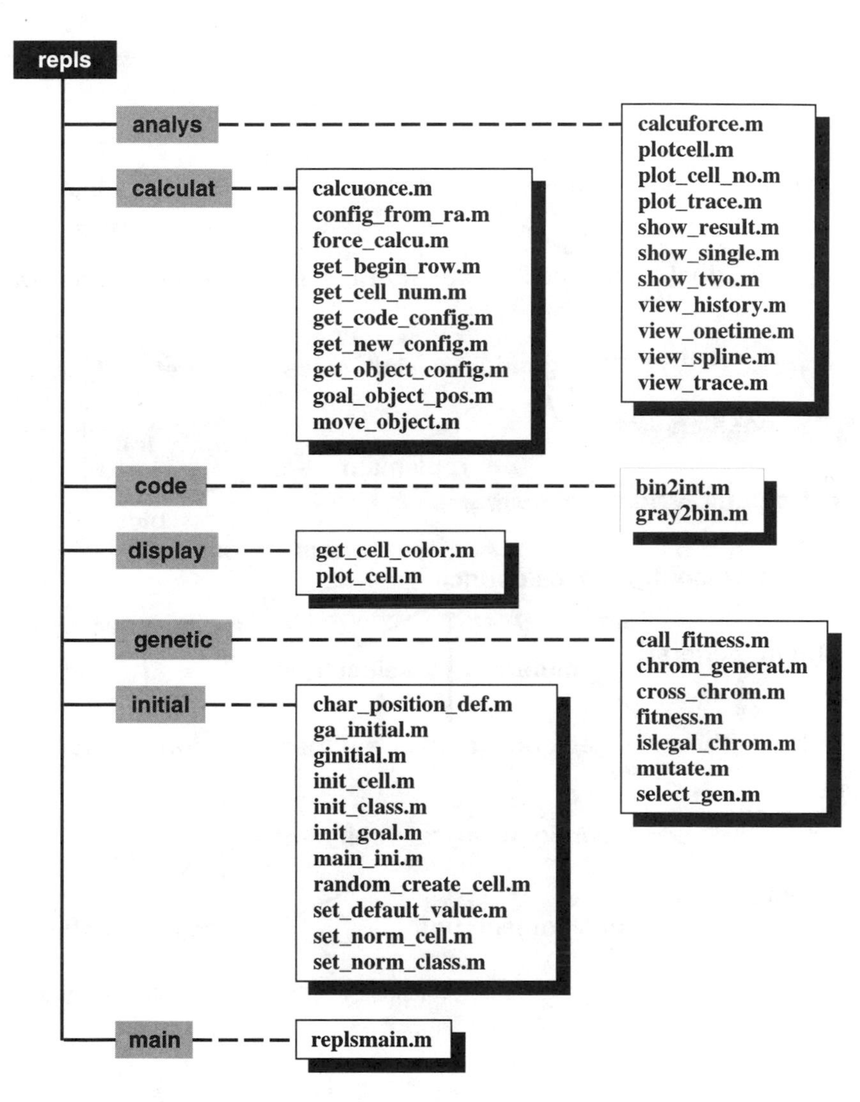

FIGURE 14.7. The file structure for the evolutionary collective box-pushing toolbox.

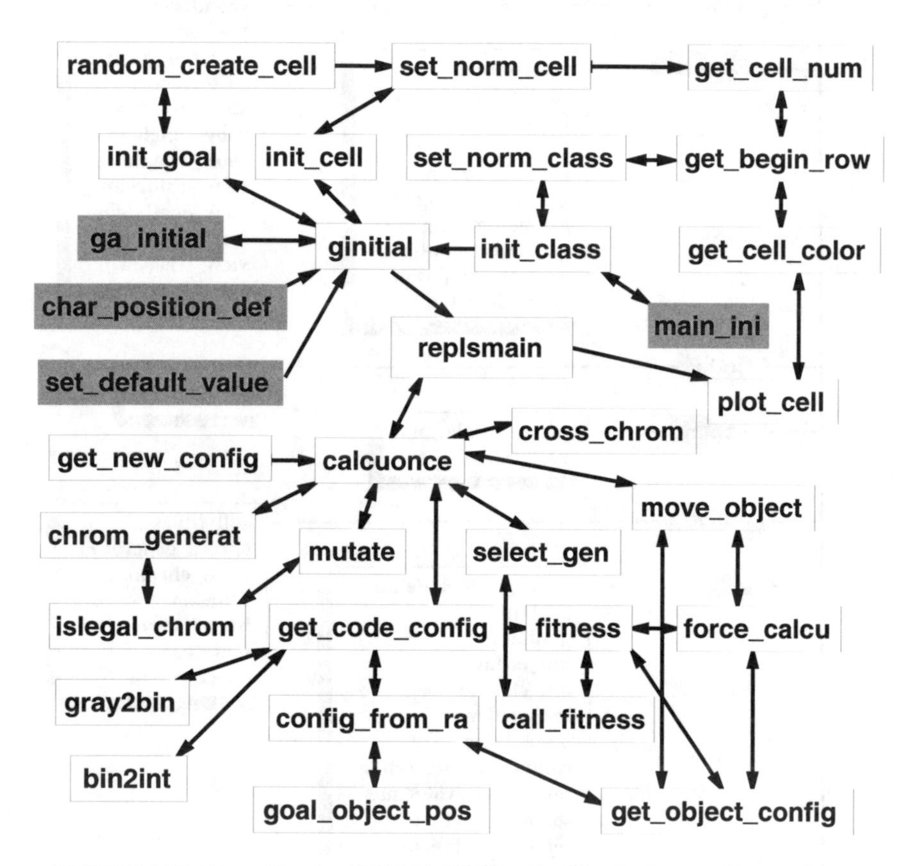

FIGURE 14.8. The calling tree for the evolutionary collective box-pushing toolbox.

Function	Description
bin2gray	convert a binary code to a gray code
bin2int	convert a binary code to an integer number
calcuonce	perform the main computation
call_fitness	calculate the fitness values for chromosomes
char_position_def	define the position of feature parameters in the matrix
chrom_generat	generate a population of chromosomes
config_from_ra	return the robot configuration from an angle
cross_chrom	mate two individuals
fitness	get the fitness function for genetic operations
force_calcu	calculate the net force on a box
ga_initial	initialize the parameters for genetic operations
get_begin_row	obtain the beginning row of `CLASS`
get_cell_color	obtain the color from `CLASS` for robot display
get_cell_num	obtain the robot number of the current class
get_code_config	convert a binary representation to a real configuration
get_new_config	obtain the new configuration of a robot
get_object_config	get the configuration of the current box
ginitial	initialize overall experimentation parameters
goal_object_pos	calculate the angle from the box and the goal
gray2bin	convert a gray code to a binary code
init_cell	initialize the robot matrix
init_class	initialize the class matrix
init_goal	initialize the goal in the experiment
int2bin	convert an integer number to a binary code
islegal_chrom	test whether a chromosome is legal
main_ini	initialize system parameters
move_object	move an object in a certain environment
mutate	mutate chromosomes at a given rate
plot_cell	plot a robot distribution in an environment
random_create_cell	create a distribution of robots randomly
replsmain	serve as the main program for pushing by repulsive forces
select_gen	select the best chromosomes for the next generation
set_default_value	assign some variables with default values
set_norm_cell	initialize the robot matrix for `CLASS`
set_norm_class	set the content of an environment for `CLASS`

TABLE 14.4. Description of the functions for collective box-pushing by artificial repulsive forces.

Function	Description
calcuforce	calculate the net force on a box
plot_cell_no	plot the current location of a robot
plot_trace	plot the trajectories of robots
plotcell	plot the current location of a robot
show_result	draw experimental results
show_single	plot the snapshots of an entire process
show_two	compare the results between two experiments
view_history	draw the history of the experiments
view_onetime	draw the distribution at a certain step
view_spline	draw the spline trajectory
view_trace	trace the motions of robots and a box

TABLE 14.5. Description of the functions for analyzing the results of box-pushing by artificial repulsive forces.

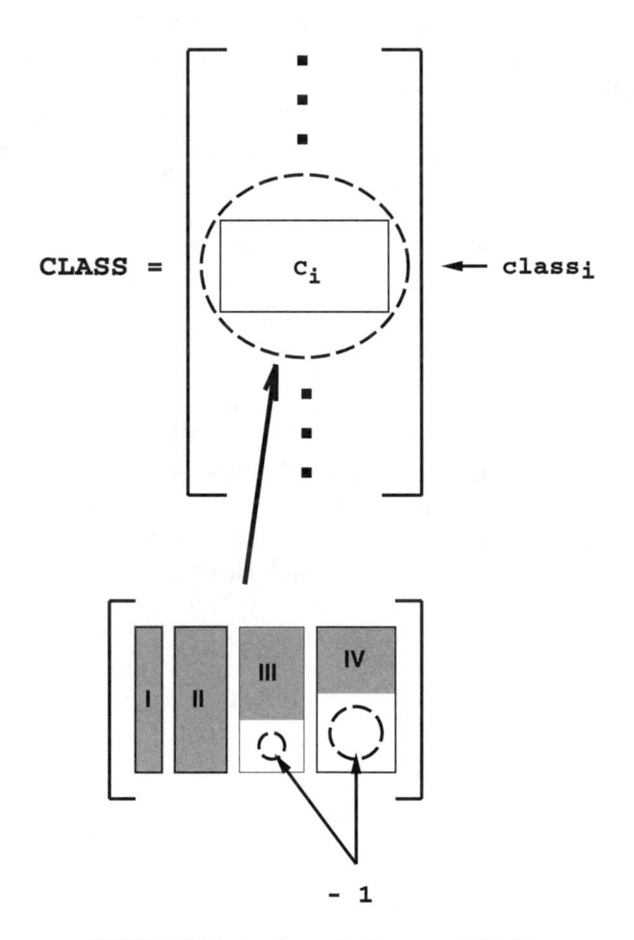

FIGURE 14.9. A schematic diagram of CLASS.

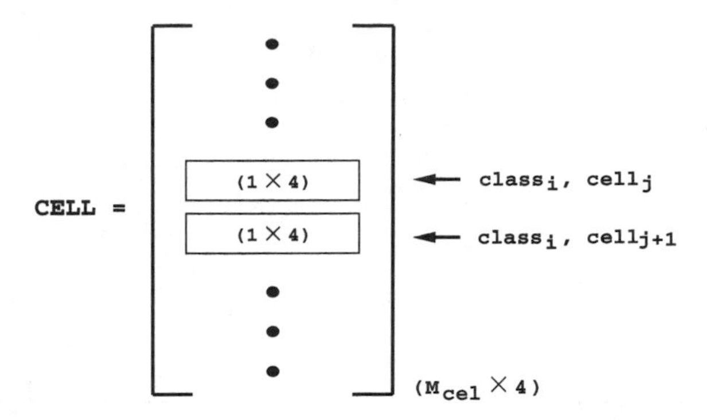

FIGURE 14.10. A schematic diagram of CELL.

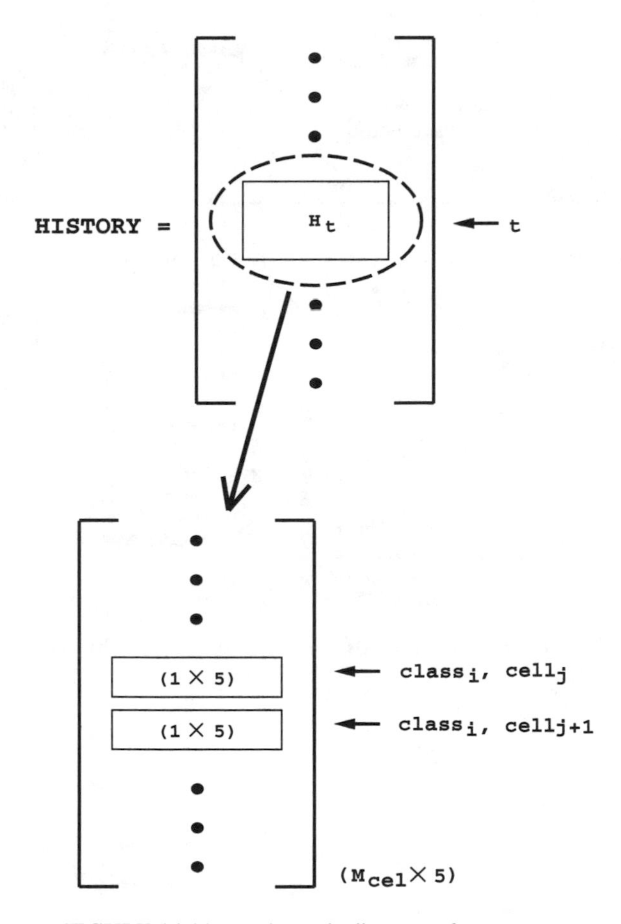

FIGURE 14.11. A schematic diagram of HISTORY.

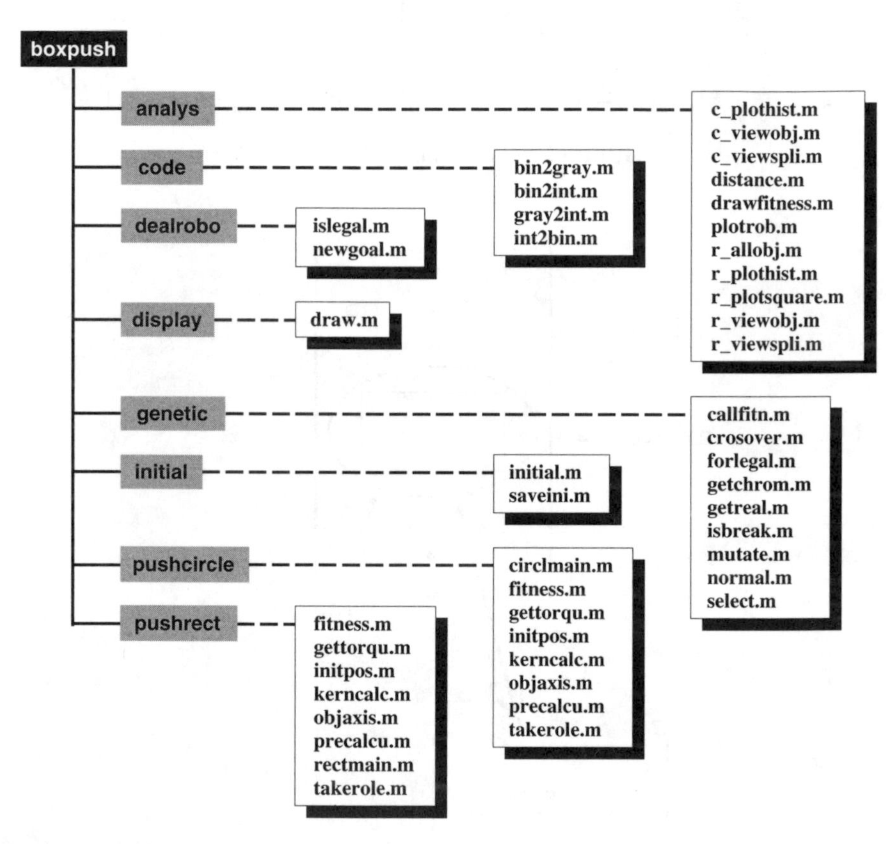

FIGURE 14.12. The file structure for the evolutionary cylindrical/cubic box-pushing toolbox.

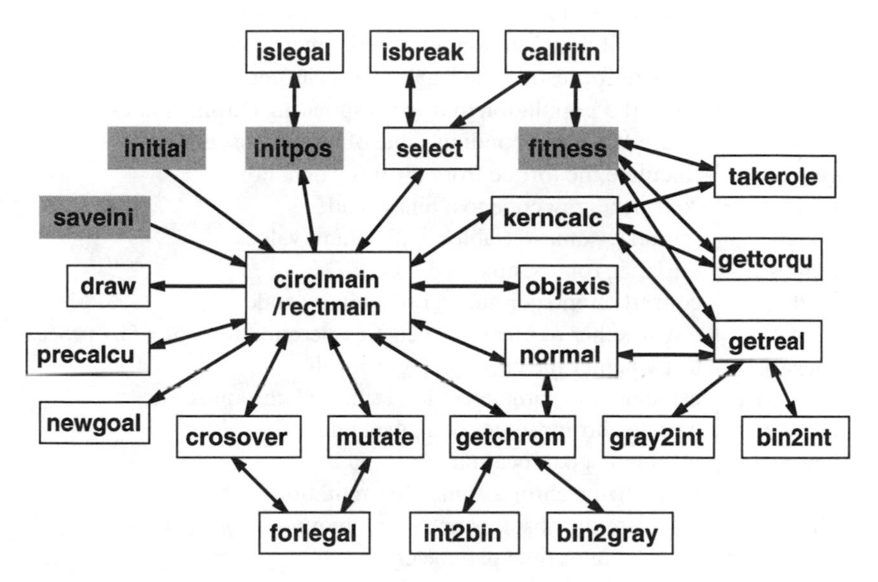

FIGURE 14.13. The calling tree for the evolutionary cylindrical/cubic box-pushing toolbox.

Function	Description
bin2gray	convert a binary code to a gray code
bin2int	convert a binary code to an integer number
callfitn	calculate the fitness values for chromosomes
circlmain	serve as the main program for cylindrical box-pushing
crosover	mate two individuals
draw	plot the results
fitness ‡	get the fitness function for genetic operations
forlegal	prepare for testing the legality of a chromosome
getchrom	convert a population to its corresponding chromosomes
getreal	return the corresponding value of a chromosome
gettorqu ‡	calculate the torque from the force on a box
gray2bin	convert a gray code to a binary code
initial	initialize some variables with default values
initpos ‡	initialize robots, box, and goal
int2bin	convert an integer number to a binary code
isbreak	test whether to break the genetic operation loop of a GA process
islegal	test whether the current case is legal
kerncalc ‡	calculate the parameters for a selected chromosome
mutate	mutate chromosomes at a given rate
newgoal	get a new goal location
normal	normalize a chromosome after mutation
objaxis ‡	calculate the absolute angle of a moving coordinate axis
precalcu ‡	precalculate some parameters
rectmain	serve as the main program for cubic box-pushing
saveini	initialize items for data storage
select	select the best chromosomes for the next generation
takerole ‡	test whether a robot is acting on a box

TABLE 14.6. Description of the functions in the toolboxes for implementing evolutionary cylindrical/cubic box-pushing behaviors. (‡: Functions in `boxpush/pushcircl/` are for cylindrical box-pushing, and those with the same function names in `boxpush/pushrect/` are for cubic box-pushing.)

Function	Description
c_plothist	plot the process of cylindrical box-pushing
c_viewobj	draw the trajectory of a cylindrical box when being pushed
c_viewspli	draw the spline trajectory of cylindrical box-pushing
distance	return the change of the distance between a box and a goal
drawfitness	plot the fitness values during the course of pushing
plotrob	draw robots
r_allobj	draw the movement of a cubic box during the course of pushing
r_plothist	plot the complete process of cubic box-pushing
r_plotsquare	draw square objects
r_viewobj	view the trajectory of a cubic box during the course of pushing
r_viewspli	view the movement trajectories of robots and a cubic box

TABLE 14.7. Description of the functions for analyzing the results of evolutionary cylindrical/cubic box-pushing.

14.4.2.3 User Configuration

initial:

1. `RobotNum` defines the number of robots in the experiments.

2. `WorkArea` defines the experimental area in which robots work. If `WorkArea` is set to $[x_{max}\ y_{max}]$, the robots will move in a rectangular area, where the coordinates of 4 corners are $(0,\ 0)$, $(0,\ y_{max})$, $(x_{max},\ y_{max})$, and $(x_{max},\ 0)$, respectively.

3. `RobotSize` has the format of [`Wid Len`]. In this toolbox, all robots are of a rectangular shape. `Wid` and `Len` define the width and length of the robots, respectively.

4. `CompareDist` is a parameter for representing the situation of a box with respect to its current goal. If the distance between the box and the goal is shorter than `CompareDist`, it is said that the box has reached the current goal.

5. `ObjectShape` specifies the shape of an object. For example, if `ObjectShape` is set to 1, it means a cylindrical box is adopted in the experiment; if it is initialized to 2, a cubic box is used.

6. `ObjectSize` defines the size of a box. In the case of cylindrical box-pushing, `ObjectSize` gives the radius of the box. In the case of cubic box-pushing, it gives the width of the box.

7. `EdgeLimit` sets a distance to keep robots away from the border of an experimental area.

8. `NewGoalLimit` limits the choices for new goal locations. In this toolbox, the distance between a current goal and a new goal must be greater than the set value.

9. `RunTimes` sets the maximum number of program running time steps.

10. `GoalVaryTm` defines a time interval to reset the location of a goal.

11. `StepRatio` is the ratio of a step size to a pushing force, i.e.,

$$d_s = \texttt{StepRatio} \cdot F, \tag{14.9}$$

where d_s is the step size, and F is the pushing force.

12. `StepChangeRatio` gives an ideal ratio for robot evolution. A box is expected to be pushed at the step that is the inverse of the distance between the box and a goal.

13. `RotateRatio` is similar to `StepRatio`. The former establishes the relationship between the amount of rotation and that of an acting torque.

14. `RegionNum` specifies the number of regions divided equally around robots. `RegionCodeLen` gives the bit length of a chromosome for representing all regions.

15. `DirectionNum` specifies the number of possible directions for a robot to move forward. `DirectCodeLen`, associated with `DirectionNum`, gives the bit length of chromosomes in genetic operations.

16. `FitnessFun` defines the name of a fitness function.

17. `BitLength` is the bit length of chromosomes in genetic operations.

18. `SingleCodeWidth` is the bit length in a chromosome for representing one robot.

19. `PopulationSize` defines the size of a population in genetic operations.

20. `TotalGenerations` indicates the number of generations in evolution for each movement of robots.

21. `CrossProb` and `MutationProb` define the probabilities of crossover and mutation in genetic operations, respectively.

22. `BreakGARatio` is used to break the genetic optimization when the specified proportion of individuals in the whole population has converged.

initpos:

1. `GoalPosition` defines the initial position of a goal in an environment.
2. `ObjectPosition` defines the position of an object. In the case of cylindrical box-pushing, it has the format of $[x\ y]$, where $(x,\ y)$ is the coordinate for the center of the box in the environment. In the case of cubic box-pushing, it includes 2 additional parameters besides the coordinate, i.e., $[x\ y\ \beta\ \gamma]$, where β is the current orientation angle of a contacting robot in the environment, and γ is the amount of accumulative rotation of the robot.

14.4.2.4 Data Structure

Assume that there are M_r robots in the experiment. There are 4 global variables to record the entire experimentation process. They are `RobotHistory`, `GoalHistory`, `ObjectHistory`, and `ChromHistory`. All of them are represented in the format as shown in Figure 14.14.

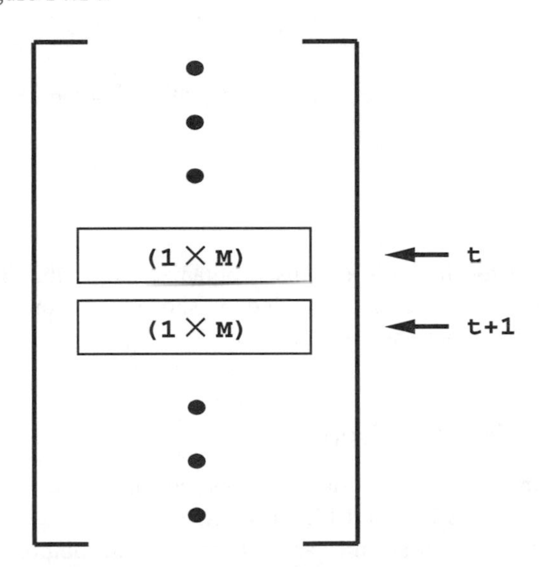

FIGURE 14.14. A schematic diagram illustrating the format of `RobotHistory`, `GoalHistory`, `ObjectHistory`, and `ChromHistory`.

`RobotHistory` records the history of robot locations. For each step, M in Figure 14.14 is equal to $2 \cdot M_r$. The vector (i.e., submatrix of size $1 \times M$ in Figure 14.14 at time t) can be expressed as follows:

$$[x_1\ y_1\ x_2\ y_2\ \cdots\ x_{M_r}\ y_{M_r}], \tag{14.10}$$

where $(x_i,\ y_i)$ is the coordinate of the ith robot.

As for `GoalHistory`, the submatrix of size $1 \times M$ in Figure 14.14 specifies the location of a goal at time t. Thus, $M = 2$. And, it has the format of $[x_g\ y_g]$, where $(x_g,\ y_g)$ denotes the current coordinate of the goal.

`ObjectHistory` records the locations of an object. In the case of cylindrical box-pushing, $M = 2$, and the submatrix has the format of $[x_o\ y_o]$. In the case of cubic box-pushing, $M = 4$, with a submatrix of $[x_o\ y_o\ \beta_o\ \gamma_o]$.

`ChromHistory` records the set of optimal individuals obtained from the evolution process. For each step, M in Figure 14.14 is equal to $2 \cdot M_r$. The vector in `ChromHistory` at time t can be expressed as follows:

$$[\mathcal{E}_1\ \mathcal{D}_1\ \mathcal{E}_2\ \mathcal{D}_2\ \cdots\ \mathcal{E}_{M_r}\ \mathcal{D}_{M_r}], \tag{14.11}$$

where $\mathcal{E}_i$ and $\mathcal{D}_i$ denote the selected region and direction for the behavior execution at time $t+1$, respectively.

14.5 Toolbox for Multi-Agent Self-Organization

14.5.1 *Architecture*

The architecture for implementing multi-agent self-organization is shown in Figure 14.15.

14.5.2 *File Structure*

The file structure for the multi-agent self-organization toolbox is shown in Figure 14.16. The calling tree of this toolbox is shown in Figure 14.17, where the functions in shaded blocks are user defined.

14.5.3 *Function Description*

The toolbox functions for multi-agent self-organization and for data processing are described in Tables 14.8 and 14.9, respectively. The functions for computing a true potential field map are the same as those in the toolbox for evolutionary multi-agent self-organization and will be described in Table 14.11.

14.5.4 *User Configuration*

User configuration for this toolbox is similar to that for the evolutionary multi-agent self-organization toolbox (see Section 14.6.4).

14.5.5 *Data Structure*

All variables have the same data structures as those in the toolbox for evolutionary multi-agent self-organization (see Section 14.6.5).

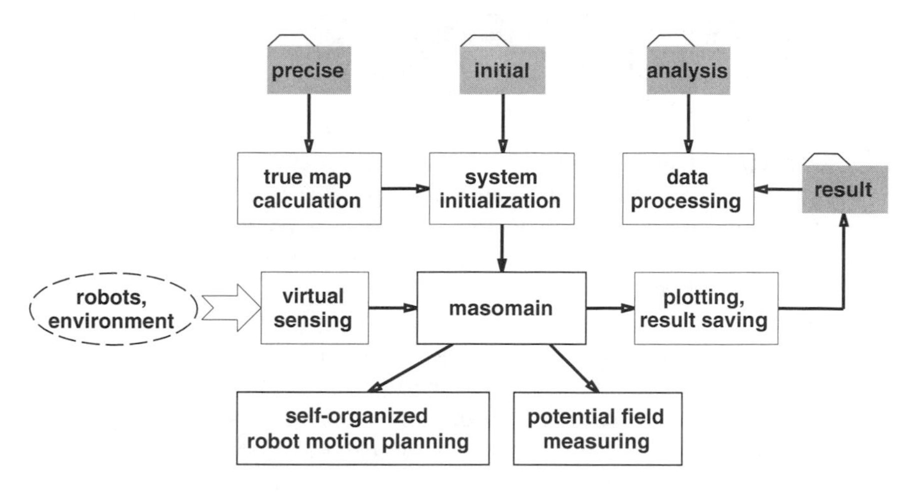

FIGURE 14.15. The architecture of the multi-agent self-organization toolbox.

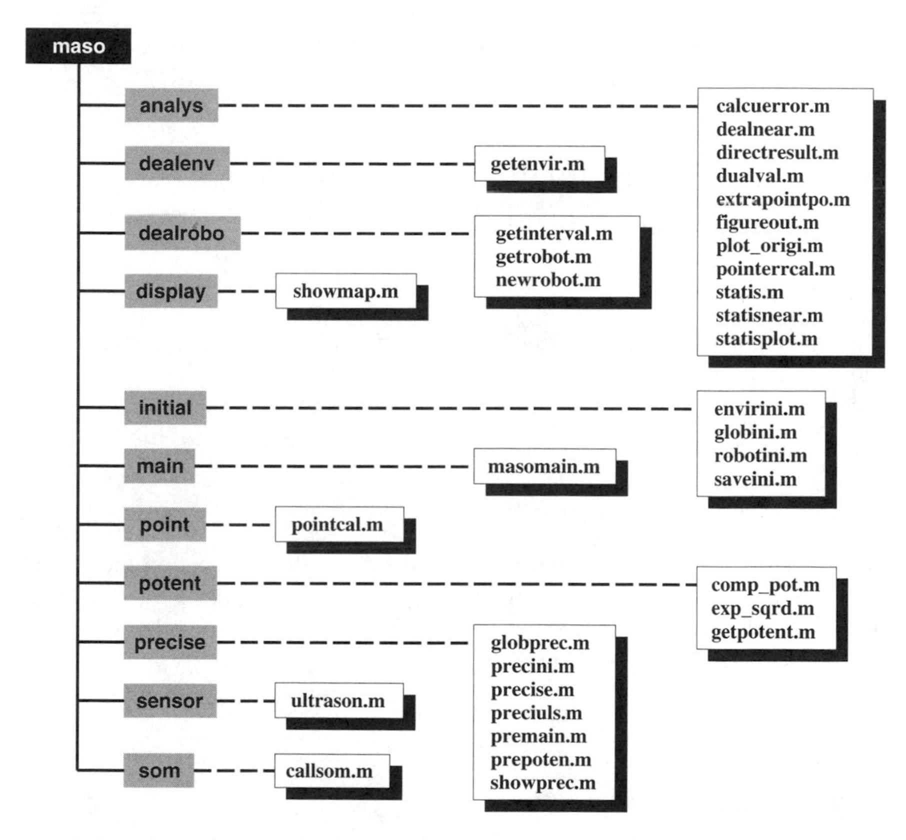

FIGURE 14.16. The file structure for the multi-agent self-organization toolbox.

Function	Description
callsom	calculate the potential field map for a designated robot
comp_pot	calculate a potential field value
envirini	initialize an environment
exp_sqrd	define a potential field function
getenvir	normalize the environment for the experiment
getinterval	calculate the interval between two robots
getpotent	perform an incremental self-organization
getrobot	set the initial positions of robots
globini	initialize some global variables
masomain	serve as the main program for self-organization
newrobot	assign new locations to robots
pointcal	calculate the potential field value for a certain location
robotini	initialize a robot group for the experiment
saveini	initialize variables for a data storage operation
showmap	display a potential field map and a robot environment
ultrason	simulate ultrasonic sensors

TABLE 14.8. Description of the functions in the evolutionary self-organization toolbox.

14.6 Toolbox for Evolutionary Multi-Agent Self-Organization

14.6.1 Architecture

The architecture for implementing evolutionary multi-agent self-organization is shown in Figure 14.18.

14.6.2 File Structure

The file structure for the evolutionary multi-agent self-organization toolbox is shown in Figure 14.19. The calling trees for this toolbox and for true potential field map building are shown in Figures 14.20 and 14.21, respectively, where the functions in shaded blocks are user defined.

14.6.3 Function Description

The toolbox functions for evolutionary self-organization, true map calculation, and data processing are described in Tables 14.10, 14.11, and 14.12, respectively.

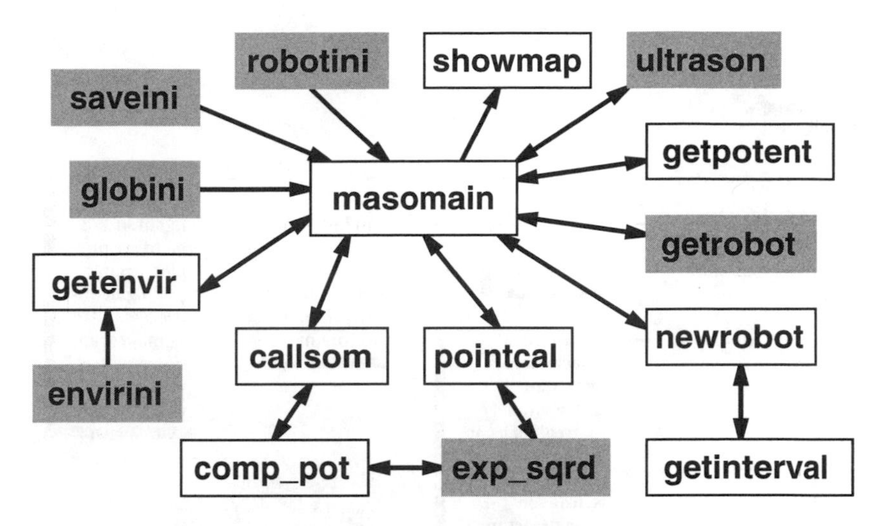

FIGURE 14.17. The calling tree for the multi-agent self-organization toolbox.

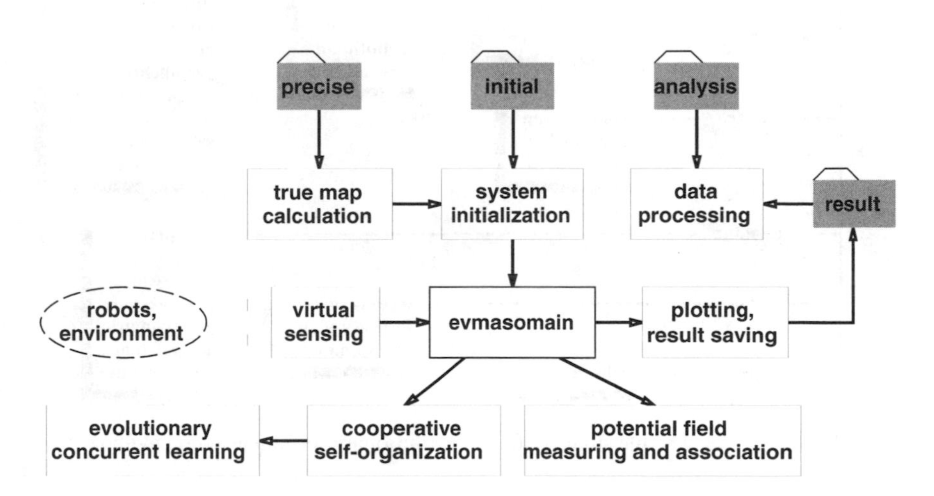

FIGURE 14.18. The architecture of the evolutionary self-organization toolbox.

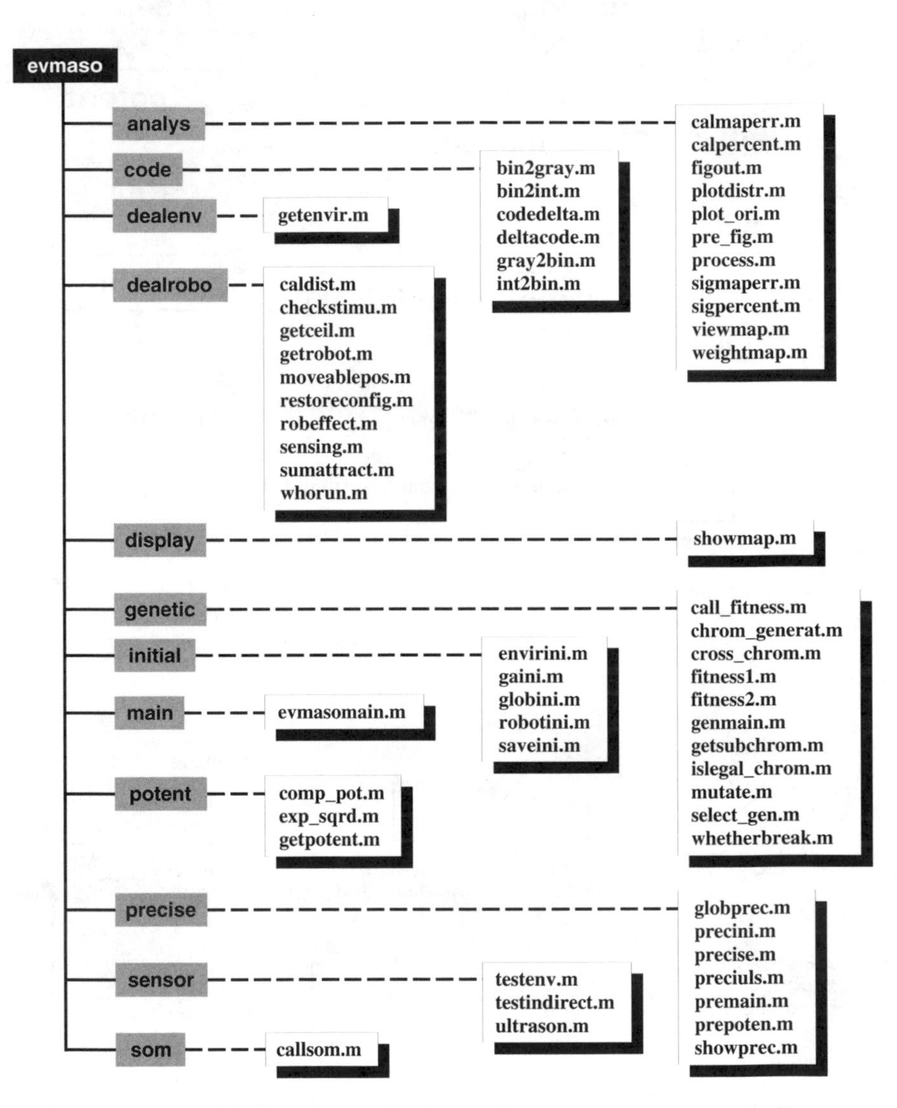

FIGURE 14.19. The file structure for the evolutionary self-organization toolbox.

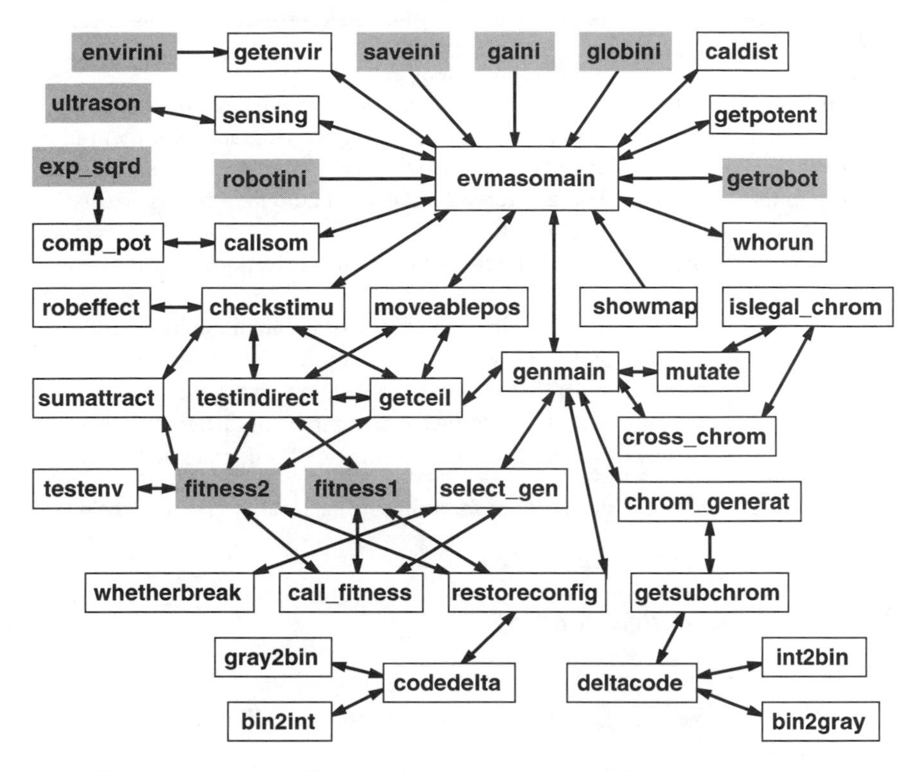

FIGURE 14.20. The calling tree for the evolutionary self-organization toolbox.

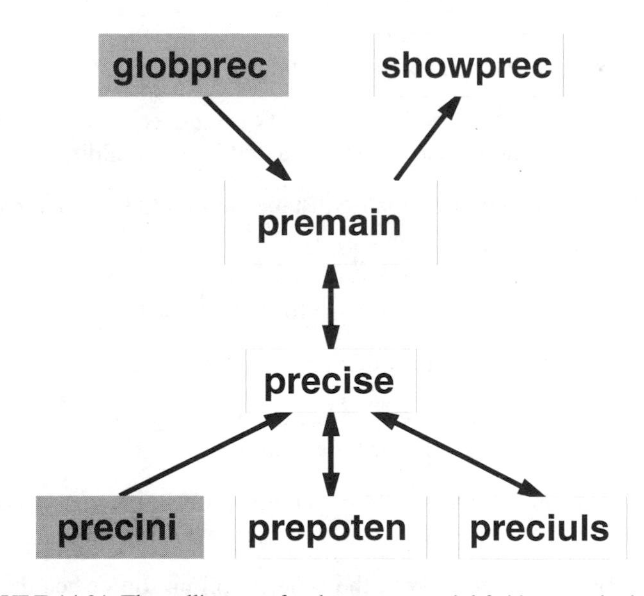

FIGURE 14.21. The calling tree for the true potential field map calculation.

Function	Description
calcuerror	calculate and display potential field errors
dealnear	count the location numbers in each behavior sector
directresult	display contour plots and robot trajectories, and save them in files
dualval	convert elements into 0 (if less than 1) or 1 (otherwise) values
extrapointpo	choose the parameters of a potential field map for comparison and save them in a file
figureout	display a potential field map and the corresponding robot trajectories
plot_origi	plot the initial distribution of robots and their environment
pointerrcal	calculate and display the error of a potential field map
statis	count the error distributions for different experiments
statisnear	compare the selectable location numbers in different experiments
statisplot	display errors in different experiments for comparison

TABLE 14.9. Description of the functions for analyzing the results of potential field map building.

14.6.4 User Configuration

envirini:

1. `EnvironName` specifies the name of a file (in the BMP format) for a robot environment.

2. `PreciseFileName` specifies a MAT file for saving a true potential field map.

3. `MaxAttitude` sets the maximum attitude of a potential field map.

4. `MapGridWidth` sets the grids in a working environment, i.e., each `MapGridWidth` point marks a grid as a unit in map building.

5. `EnvThresh` specifies a geometric distance threshold for robot interaction.

gaini:

1. `FitnessFun` specifies the file name for fitness functions in a string matrix.

2. `SingleCodeWidth` sets the bit length in a chromosome for a single robot.

3. `PopulationSize` and `GenerationSize` assign the sizes of population and generation, respectively, for genetic optimization.

4. `CrossProb` and `MutationProb` set the probabilities of crossover and mutation, respectively.

5. `BreakRatio` is used to break the genetic optimization when the specified proportion of individuals in the whole population has converged.

globini:

1. `FreeIndex`, `ObstacleIndex`, and `RobotIndex` must be set to different numbers for object identification at a certain location.

2. `PotentFunc` specifies the name of the function that defines a potential field expression.

3. `MaxRunStep` sets the maximum number of loops in computation. The program can break the loops before the running time step reaches `MaxRunStep` if it satisfies a convergence condition. Otherwise, the computation will be stopped at `MaxRunStep`.

robotini:

1. `RobotNum` sets the total number of autonomous robots in a group. `SensorNum` sets the total number of sensors mounted on each robot.

2. `SensorType` gives the name of the function that defines the sensing ability of a robot.

3. `IncrementWeight` sets the maximum increment for the best selection.

4. `MaxMoveStep` defines the maximum movement step of each robot.

saveini:

1. `SaveDirectory` specifies the directory name with a full path for saving a data file (in the MAT format).

2. `SaveItem` assigns all the global variables required to be stored for future analysis.

14.6.5 Data Structure

Assume that there are M cooperative robots in an environment of size $X_0 \times Y_0$ for evolutionary potential field map building. An example of the environment is shown in Figure 14.22. Matrix `RealEnviron` is used to specify this environment, as follows:

$$\texttt{RealEnviron} = \begin{bmatrix} \ddots & \vdots & \\ \cdots & \mathrm{E}_{ij} & \cdots \\ & \vdots & \ddots \end{bmatrix}_{(Y_0 \times X_0)}, \tag{14.12}$$

where

$$\mathrm{E}_{ij} = \begin{cases} \texttt{FreeIndex}, & \text{if grid } (j,i) \text{ belongs to working space,} \\ \texttt{ObstacleIndex}, & \text{if grid } (j,i) \text{ belongs to an obstacle.} \end{cases} \tag{14.13}$$

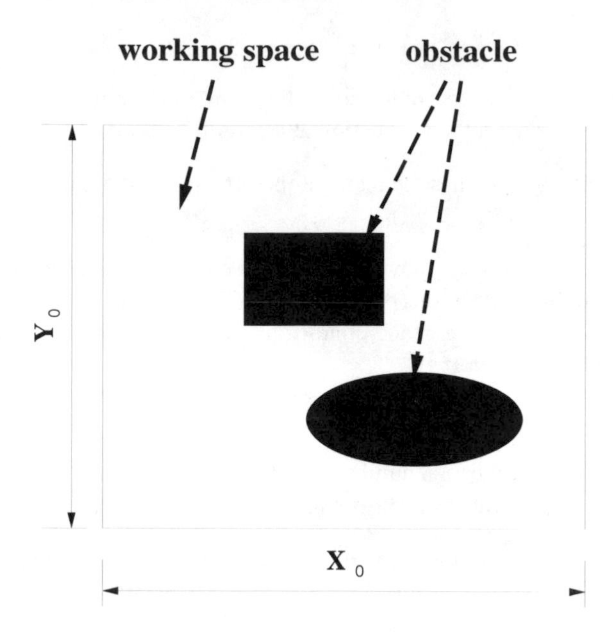

FIGURE 14.22. A robot environment.

In the design of this toolbox, variable `MapGridWidth` is used to set the spacing for robot map building, as illustrated in Figure 14.23. Thus, the size of a map will be $X \times Y$, where

$$X = \lfloor \frac{X_0}{\texttt{MapGridWidth}} \rfloor, \quad Y = \lfloor \frac{Y_0}{\texttt{MapGridWidth}} \rfloor, \tag{14.14}$$

where $\lfloor \cdot \rfloor$ converts a real number into an integer.

Matrix `VisitTimes` records the total number of robots that have visited a certain location. It can be expressed as follows:

$$\texttt{VisitTimes} = \begin{bmatrix} \ddots & \vdots & \\ \cdots & \mathrm{N}_{ij} & \cdots \\ & \vdots & \ddots \end{bmatrix}_{(Y \times X)}, \tag{14.15}$$

where $\mathrm{N}_{ij} \in \{0,\ 1,\ 2,\ \cdots\}$.

`VisitPlusEnviron` marks the locations visited by robots based on `RealEnviron`, i.e.,

$$\texttt{VisitPlusEnviron} = \begin{bmatrix} \ddots & \vdots & \\ \cdots & V_{ij} & \cdots \\ & \vdots & \ddots \end{bmatrix}_{(Y_0 \times X_0)}, \tag{14.16}$$

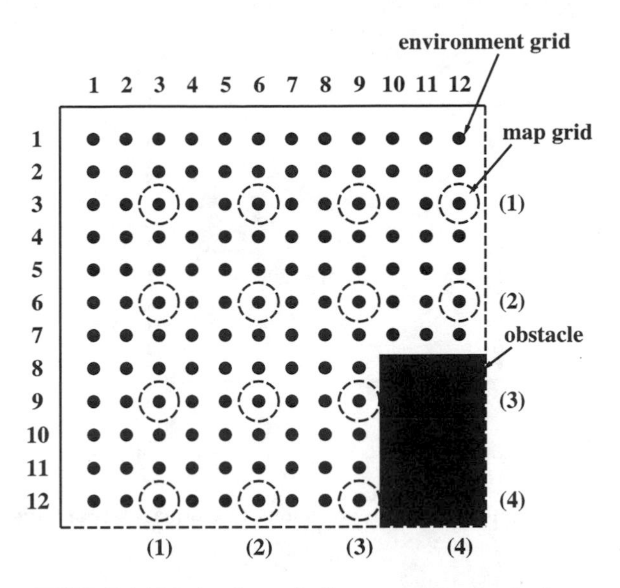

FIGURE 14.23. A schematic diagram of the map lattice.

where

$$V_{ij} = \begin{cases} \texttt{FreeIndex}, & \text{if grid } (j,i) \text{ belongs to working space,} \\ \texttt{ObstacleIndex}, & \text{if grid } (j,i) \text{ belongs to an obstacle,} \\ \texttt{MarkColorIndex}, & \text{if grid } (j,i) \text{ is visited by robot(s),} \end{cases} \tag{14.17}$$

where `MarkColorIndex` is the 8-bit complementary number of `FreeIndex`.

`PrecisePotent` stores the true potential field value at each map grid, i.e.,

$$\texttt{PrecisePotent} = \begin{bmatrix} \ddots & \vdots & \\ \cdots & \bar{\mathcal{U}}_{ij} & \cdots \\ & \vdots & \ddots \end{bmatrix}_{(Y \times X)}, \tag{14.18}$$

where $\bar{\mathcal{U}}_{ij}$ is the true potential field value at (j, i).

The measurements or associations of a temporary potential field map at any time for any locations are stored in a three-dimensional data structure, `TempPotentField`. The corresponding probabilities are in `TempPotentWeight`. They both have the format of Figure 14.24, in which the longitudinal array at grid (j, i) has been highlighted. The array of temporary probabilities for the specific grid can be expressed as follows:

$$\Phi_{ij}^{t} = \{w_{ij}^{t_1}, w_{ij}^{t_2}, \cdots, w_{ij}^{t_k}\}. \tag{14.19}$$

Similarly, the array of temporary potential field values for the specific grid can be written as follows:

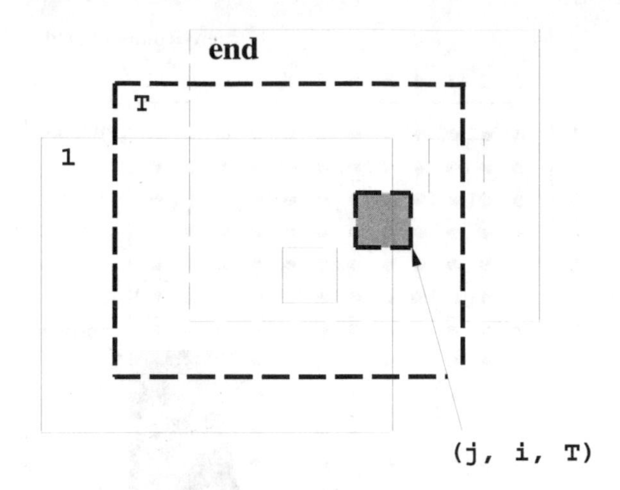

FIGURE 14.24. A schematic diagram illustrating the format of `TempPotentWeight` and `TempPotentField`.

$$\Omega_{ij}^{t} = \{\mathcal{U}_{ij}^{t_1}, \mathcal{U}_{ij}^{t_2}, \cdots, \mathcal{U}_{ij}^{t_k}\}, \tag{14.20}$$

where $w_{ij}^{t_k}$ indicates the probability for $\mathcal{U}_{ij}^{t_k}$ at time t_k.

Matrix `TotalWeight` records the history of weight changing in behavior selection, as shown in Figure 14.25. Each row in `TotalWeight` represents the set of weights corresponding to all primitive behaviors, except the first element, which indicates a behavior code selected in response to the stimulus at time t. W^t can be expressed as follows:

$$W^{t} = \{B_k, \omega_1^t, \omega_2^t, \cdots, \omega_K^t\}, \tag{14.21}$$

where B_k is the code for the kth primitive behavior, and

$$\sum_{i=1}^{K} \omega_i^t = 1. \tag{14.22}$$

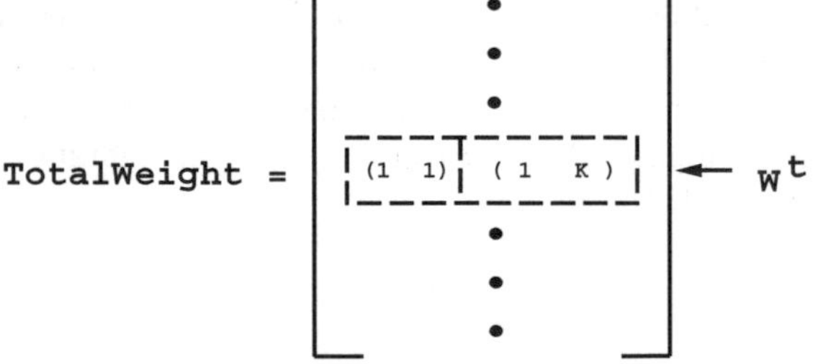

TotalWeight = (1 1) (1 K) ← W^t

FIGURE 14.25. A schematic diagram of `TotalWeight`.

Matrix AllRobotConfig, as shown in Figure 14.26, stores the history of robot configurations. At time t, submatrix C^t is added to the bottom of AllRobotConfig where

$$C^t = \begin{bmatrix} \vdots & \vdots \\ y_k & x_k \\ \vdots & \vdots \end{bmatrix}_{(M \times 2)}, \tag{14.23}$$

where (x_k, y_k) specifies the location of robot k at time t.

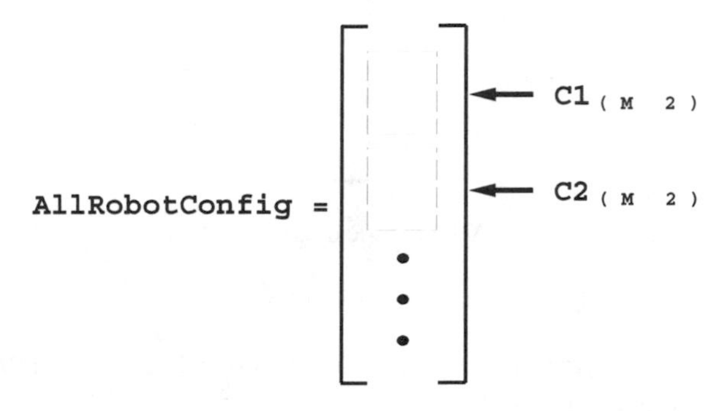

FIGURE 14.26. A schematic diagram of AllRobotConfig.

TotalPotent uses the similar rule, as shown in Figure 14.27, to store all temporary potential field map measurements corresponding to mapping grids. At time t, submatrix U^t can be written as follows:

$$U^t = \begin{bmatrix} \ddots & \vdots & \\ \cdots & \mathcal{U}_{ij}^t & \cdots \\ & \vdots & \ddots \end{bmatrix}_{(Y \times X)}, \tag{14.24}$$

where $\mathcal{U}_{ij}^t$ is the potential field value at (j, i) at time t.

14.7 Example

Let us take a look at an example of how to carry out a multi-robot experiment on evolutionary multi-agent self-organization using the provided toolbox. The parameters for this example are given in Table 14.13.

14.7.1 True Map Calculation

First of all, we initialize several true map calculation functions, according to the parameters in Table 14.13.

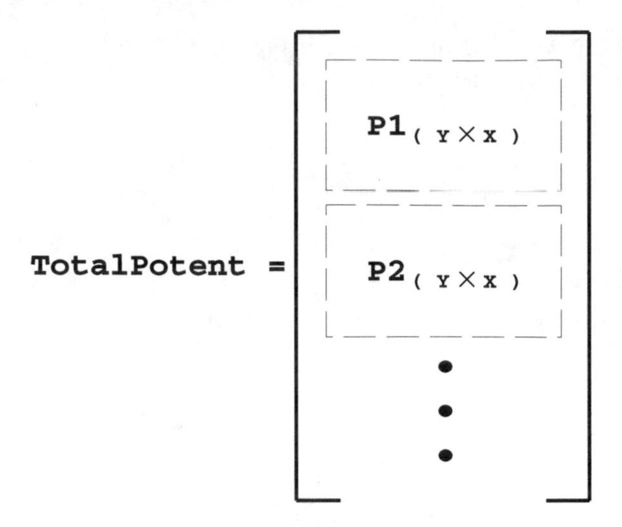

FIGURE 14.27. A schematic diagram of TotalPotent.

1. evmaso/precise/glibprec.m

   In this function, variables SensorNum, MapGridWidth, and EnvironName are initialized as follows:

   ```
   SensorNum=16;
   MapGridWidth=5;
   EnvironName='testimg.bmp';
   ```

   where the simulated environment file, testimg.bmp, is kept in the same directory, or the path name for the MATLAB environment is set to share the directory that contains this file.

2. evmaso/precise/premain.m

   This file assigns a file for saving a true potential field map.

   ```
   save precis PrecisePotent;
   ```

3. evmaso/precise/precini.m

   This file assigns two function names for potential field function PrecisePotentFunc and sensor feature description PreciseSensor, respectively:

   ```
   PrecisePotentFunc='prepoten';
   PreciseSensor='preciuls';
   ```

   Thereafter, two named functions, prepoten.m and preciuls.m, need to be defined.

4. `evmaso/precise/prepoten.m`

   This file implements the function for calculating a potential field value based on the equation given in Table 14.13:

```
function DirectPotent=prepoten(DirectDist)
% PREPOTEN Define the potential field function.

%    Copyright (c) 1998-2001 Jiming Liu & Jianbing Wu

DirectPotent=exp(-DirectDist/5);
```

5. `evmaso/precise/preciuls.m`

   As far as the sensor feature description is concerned, `evmaso/precise/preciuls.m` can be defined in the same way as `evmaso/precise/prepoten.m`.

Now, `evmaso/precise/` will contain the following files:

```
sccs5:~/evmaso/precise>ls
globprec.m    precise.m    premain.m    showprec.m
precini.m     preciuls.m   prepoten.m   testimg.bmp
sccs5:~/evmaso/precise>_
```

After running `evmaso/precise/premain.m`, e.g.,

```
>> cd evmaso/precise

>> premain

>>
```

the MAT file, `precis.mat`, with a true potential field map will be created in the current directory, which now contains the following files:

```
sccs5:~/evmaso/precise>ls
globprec.m   precis.mat   preciuls.m   prepoten.m   testimg.bmp
precini.m    precise.m    premain.m    showprec.m
sccs5:~/evmaso/precise>_
```

14.7.2 Initialization

Next, we initialize the functions for evolutionary multi-agent self-organization, using the parameters in Table 14.13.

1. `evmaso/initial/globini.m`

   In this function, potential field function `PotentFunc` and variable `MaxRunStep` are initialized as follows:

```
PotentFunc='exp_sqrd';

MaxRunStep=120;
```

where exp_sqrd.m is created for potential field measurements. This function is the same as evmaso/precise/prepoten.m given above.

2. evmaso/initial/envirini.m

   This file initializes the following variables with respect to a simulated environment:

```
PreciseFileName='precis';

EnvironName='testimg.bmp';

MaxAttitude=9;

MapGridWidth=5;

EnvThresh=3;
```

   where precis.mat and testimg.bmp are put in a directory set in PATH.

3. evmaso/initial/robotini.m

   In this file, the following variables are initialized:

```
RobotNum=6;

SensorType='ultrason';

SensorNum=16;

IncrementWeight=0.2;

MaxMoveStep=7;
```

   As associated with SensorType, ultrason.m is created to specify the features of the sensors.

4. evmaso/initial/getrobot.m

   In this file, the initial configurations of N robots are defined:

```
RobotConfig=[20 30;
              5  5;
             20 10;
             10 40;
```

```
45 25;
55  5];
```

5. `evmaso/initial/gaini.m`

   In this file, several parameters for genetic operations are defined:

```
FitnessFun=['fitness1';'fitness2'];
SingleCodeWidth=8;
PopulationSize=[20 30 45 65 90 120];
GenerationSize=[ 8 12 18 26 36  48];
CrossProb=0.6;
MutationProb=[0.1 0.05 0.005];
BreakRatio=0.8;
```

   `fitness1.m` and `fitness2.m` give two example fitness functions for our experiments. They are created before starting the main program.

6. `evmaso/initial/gaini.m`

   Through the following steps, the given variables will be saved in `evmaso/result/exper.mat`:

```
SaveItem='TotalPotent AllRobotConfig VisitTimes MapSize RunTimes';
SaveItem=[SaveItem ' TotalRunningRobot PrecisePotent VisitPlusEnviron'];
SaveItem=[SaveItem ' MapGridWidth RobotNum SensorNum DirectionWeight'];
SaveItem=[SaveItem ' TempPotentField TempPotentWeight TotalWeight'];
SaveItem=[SaveItem ' BeginTime EndTime'];

SaveDirectory='../result/exper';
```

14.7.3 Start-Up

Before we run the main program, we will set the path by creating file `evmaso/run/startup.m`, as follows:

```
path(path,'../evmaso/code');
path(path,'../evmaso/dealenv');
path(path,'../evmaso/dealrobo');
path(path,'../evmaso/display');
path(path,'../evmaso/genetic');
path(path,'../evmaso/initial');
path(path,'../evmaso/main');
path(path,'../evmaso/potent');
path(path,'../evmaso/precise');
path(path,'../evmaso/sensor');
path(path,'../evmaso/som');
```

Now, we can run the main program and begin the experiment on evolutionary multi-agent self-organization as follows:

```
>> cd evmaso/run

>> startup

>> evmasomain

>>
```

Thereafter, file `exper.mat` will be saved in `evmaso/result/`.

14.7.4 *Result Display*

As an example, function `/evmaso/analys/plotdistr.m` can be called to display the visited locations in the simulated environment. The resulting MAT file is designated as follows:

```
FileName='exper';
```

After the following operations,

```
>> cd evmaso/analys

>> plotdistr

>>
```

a window will pop up in which the experimental result is displayed, as shown in Figure 14.28, and a postscript file, `exper_distr.ps`, will be saved in `/evmaso/result/`.

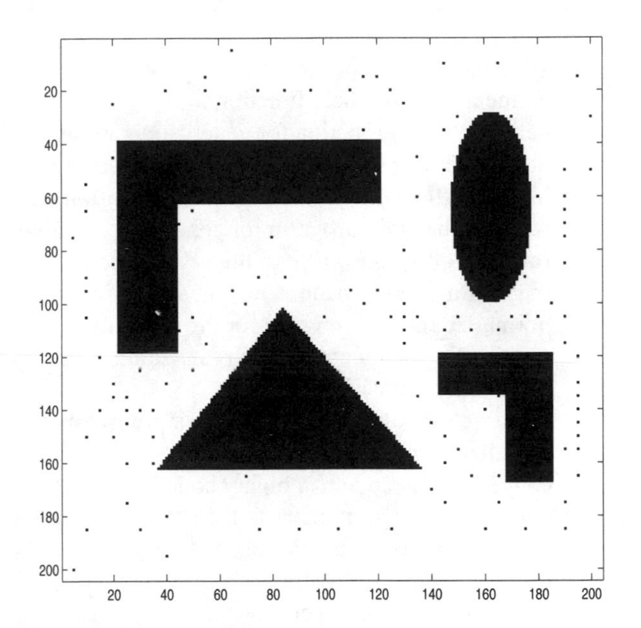

FIGURE 14.28. A graphical display in the example (©1999 IEEE).

Function	Description
bin2gray	convert a binary code to a gray code
bin2int	convert a binary code to an integer number
caldist	calculate the distance between two robots
callsom	calculate thc potential field map for a designated robot
call_fitness	calculate the fitness value for a population of chromosomes
checkstimu	check the stimulus for a running robot
chrom_generat	generate a population of chromosomes
comp_pot	calculate a potential field value
cross_chrom	perform a crossover operation
deltacode	convert a robot configuration to a gray code representation
deltaconfig	convert a gray code representation to a robot configuration
envirini	initialize an environment
evmasomain	serve as the main program for evolutionary self-organization
exp_sqrd	define a potential field function
fitness1	define a fitness function for genetic operations
fitness2	define a fitness function for genetic operations
gaini	initialize the parameters for genetic operations
genmain	serve as the main program for genetic algorithm operations
getceil	round a robot coordinate to the nearest location toward the origin of an environment
getenvir	normalize the environment for the experiment
getpotent	perform an incremental self-organization
getrobot	set the initial positions of robots
getsubchrom	generate the subchromosomes for a group robot
globini	initialize some global variables
gray2bin	convert a gray code to a binary code
int2bin	convert an integer number to the corresponding binary code
islegal_chrom	test the legality of individuals in a population
moveablepos	determine feasible positions for robots
mutate	perform a mutation operation
restoreconfig	perform a transformation from chromosomes
robeffect	get the interaction between two robots
robotini	initialize robots and their sensors
saveini	initialize variables for a data storage operation
select_gen	perform a selection operation
sensing	sense an environment by each selected robot
showmap	display a potential field map and a robot environment
sumattract	return the sum of attraction from other robots
testenv	sense an environment in a certain direction
testindirect	determine the distance between two robots of different groups
ultrason	simulate ultrasonic sensors
whetherbreak	test whether to stop genetic operations
whorun	determine executable robots at time t

TABLE 14.10. Description of the functions in the evolutionary self-organization toolbox.

Function	Description
globprec	initialize true potential field map calculation
precini	initialize sensor simulation and a potential field function
precise	calculate the potential field distribution for an environment
preciuls	perform a sensing operation for true map calculation
premain	serve as the main program for computing a true potential field map
prepoten	define a potential field function
showprec	display a true potential field map

TABLE 14.11. Description of the functions for computing true potential field values.

Function	Description
calmaperr	calculate and display the errors of a built map
calpercent	calculate the percentage of a reached area
figout	display the trajectories of robots and save them in files
plot_ori	display an initial state
plotdistr	mark the positions of robots in an environment
pre_fig	display comparable robot trajectories and save them in files
process	view the history of robot movements
sigmaperr	calculate the second-moment error in map building
sigpercent	calculate the percentage of an unreached area
viewmap	view the potential field distribution at time t
weightmap	plot the weight matrix

TABLE 14.12. Description of the functions for analyzing the results of potential field map building.

Parameter	**Value**
number of robots	6
sensory section	16
simulated environment	`testimg.bmp`
`MapGridWidth`	5
potential field function	$\Sigma exp(-d_i/5)$
true map file	`precis.mat`
maximum number of running time steps	120
potential function	`exp_sqrd.m`
sensor feature description	`ultrason.m`
maximum movement step	7
behavior vector increment	0.2
bitlength for each robot in GAs	8
population size	20/30/45/65/90/120
generations per step	8/12/18/26/36/48
crossover probability	0.6
mutation probability	0.1/0.05/0.005
`BreakRatio`	0.8

TABLE 14.13. Parameters as used in the experiments.

References

[AB97] R. C. Arkin and T. Balch. AuRA: Principles and practice in review. *Journal of Experimental and Theoretical Artificial Intelligence*, 9(2-3):175–189, 1997.

[ACF+98] R. Alami, R. Chatila, S. Fleury, M. Ghallab, and F. Ingrand. An architecture for autonomy. *International Journal of Robotics Research – Special Issue on Integrated Architectures for Robot Control and Programming*, 17(4):315–337, 1998.

[AFH+98] R. Alami, S. Fleury, M. Herrb, F. Ingrand, and F. Robert. Multi-robot cooperation in the Martha project. *IEEE Robotics and Automation Magazine – Special Issue on Robotics and Automation in the European Union*, 5(1):36–47, 1998.

[AG96] R. Ahmed and D. Gray. Immunological memory and protective immunity: Understanding their relation. *Science*, 272:54–60, 1996.

[AMI89] H. Asama, A. Matsumoto, and Y. Ishida. Design of an autonomous and distributed robot system: ACTRESS. In *Proceedings of the IEEE/RSJ International Workshop on Intelligent Robots and Systems*, pages 283–290, Tsukuba, 1989.

[Ang93] P. J. Angeline. Evolutionary Algorithms and Emergent Intelligence. Ph.D. Thesis, Ohio State University, 1993.

[AOS89] T. Arai, H. Ogata, and T. Suzuki. Collision avoidance among multiple robots using virtual impedance. In *Proceedings of the IEEE/RSJ International Workshop on Intelligent Robots and Systems*, pages 479–485, Tsukuba, 1989.

[Ark89] R. C. Arkin. Motor schema-based mobile robot navigation. *International Journal of Robotics Research*, 8(4):92–112, 1989.

[Ark98] R. C. Arkin. *Behavior-Based Robotics*. The MIT Press, Cambridge, MA, 1998.

[BA95] T. Balch and R. C. Arkin. Motor schema-based formation control for multiagent robot teams. In *Proceedings of the First International Conference on Multi-Agent Systems*, pages 10–16, AAAI Press, Menlo Park, 1995.

[BA98] T. Balch and R. C. Arkin. Behavior-based formation control for multi-robot teams. *IEEE Transactions on Robotics and Automation*, 14(6):1–15, 1998.

[BA00] S. C. Botelho and R. Alami. A multi-robot cooperative task achievement system. In *Proceedings of the IEEE International Conference on Robotics and Automation*, pages 2716–2721, San Francisco, 2000.

[Bal94] C. Balkenius. Biological Learning and Artificial Intelligence. Technical Report, Cognitive Science Department, Lund University, Sweden, 1994.

[Bal97] T. Balch. Learning roles: Behavioral diversity in robot teams. In *Proceedings of the AAAI-97 Workshop on Multiagent Learning*, pages 7–12, Providence, RI, 1997.

[Bal98] T. Balch. Behavioral Diversity in Learning Robot Teams. Ph.D. Thesis, College of Computing, Georgia Institute of Technology, 1998.

[BB99] H. M. Botee and E. W. Bonabeau. Evolving ant colony optimization. *Advances in Complex Systems*, 1(2-3):149–159, 1999.

[BBHCD96] S. Betge-Brezetz, P. Hebert, R. Chatila, and M. Devy. Uncertain map making in natural environments. In *Proceedings of the IEEE International Conference on Robotics and Automation*, pages 1048–1053, Minneapolis, 1996.

[BD97] A. Billard and K. Dautenhahn. Grounding communication in situated, social robots. In *Proceedings of TIMR97, Towards Intelligent Mobile Robots Conference*, Manchester, UK, 1997.

[BDT99] E. W. Bonabeau, M. Dorigo, and G. Theraulaz. *Swarm Intelligence: From Natural to Artificial Systems.* Oxford University Press, New York, 1999.

[Bed92] M. A. Bedau. Philosophical aspects of Artificial Life. In F. J. Varela and P. Bourgine, editors, *Towards a Practice of Autonomous Systems: Proceedings of the First European Conference on Artificial Life*, pages 494–503, The MIT Press/Bradford Books, Cambridge, MA, 1992.

[Ben88] G. Beni. The concept of cellular robotic systems. In *Proceedings of the IEEE International Symposium on Intelligent Control*, pages 57–62, Arlington, VA, 1988.

[BH94] A. Baader and G. Hirzinger. A self-organizing algorithm for multisensory surface reconstruction. In *Proceedings of the IEEE/RSJ/GI International Conference on Intelligent Robots and Systems*, pages 81–88, Munich, 1994.

[BH00] T. Balch and M. Hybinette. Social potentials for scalable multirobot formations. In *Proceedings of the IEEE International Conference on Robotics and Automation*, San Francisco, 2000.

[BHD94] R. Beckers, O. Holland, and J. Deneubourg. From local actions to global tasks: Stigmergy and collective robotics. In R. A. Brooks and P. Maes, editors, *Proceedings of the Fourth International Conference on Artificial Life*, pages 181–189, The MIT Press, Cambridge, MA, 1994.

[BHS97] T. Bäck, U. Hammel, and H.-P. Schwefel. Evolutionary computation: Comments on the history and current state. *IEEE Transaction on Evolutionary Computation*, 1(1):3–17, 1997.

[BIM00] A. Billard, A.-J. Ijspeert, and A. Martinoli. A multi-robot system for adaptive exploration of a fast changing environment: Probabilistic modeling and experimental study. *Connection Science*, 11(3-4):357–377, 2000.

[BK98] O. Brock and O. Khatib. Executing motion plans for robots with many degrees of freedom in dynamic environment. In *Proceedings of the IEEE International Conference on Robotics and Automation*, pages 1–6, Leuven, Belgium, 1998.

[BMF+00] W. Burgard, M. Moors, D. Fox, R. Simmons, and S. Thrun. Collaborative multi-robot exploration. In *Proceedings of the IEEE International Conference on Robotics and Automation*, San Francisco, 2000.

[Bon98] E. W. Bonabeau. Social insect colonies as complex adaptive systems. *Ecosystems*, 1(5):437–443, 1998.

[Bre62] H. J. Bremermann. Optimization through evolution and recombination. In M. C. Yovits et al., editors, *Self-Organizing Systems*, pages 93–106, Spartan Books, Washington, 1962.

[Bro86] R. A. Brooks. A robust layered control system for a mobile robot. *IEEE Journal of Robotics and Automation*, RA-2(1):14–23, 1986.

[Bro91] R. A. Brooks. Intelligence without Reason. Memo 1293, MIT AI Lab, USA, 1991.

[Bro92] R. A. Brooks. Artificial Life and real robots. In F. J. Varela and P. Bourgine, editors, *Towards a Practice of Autonomous Systems: Proceedings of the First European Conference on Artificial Life*, pages 3–10, The MIT Press/Bradford Books, Cambridge, MA, 1992.

[Bro99] R. A. Brooks. *Cambrian Intelligence: the Early History of New AI*. The MIT Press, Cambridge, MA, 1999.

[BRS94] M. Betke, R. L. Rivest, and M. Singh. Piecemeal Learning of an Unknown Environment. Memo 1474, MIT AI Lab, USA, 1994.

[BT95] E. W. Bonabeau and G. Theraulaz. Why do we need Artificial Life? In C. G. Langton, editor, *Artificial Life: An Overview*, pages 303–326, The MIT Press, Cambridge, MA, 1995.

[BTC98] E. W. Bonabeau, G. Theraulaz, and F. Cogne. The Design of Complex Architectures by Simple Agents. Working Paper 98-01-005, SFI, USA, 1998.

[BW89] G. Beni and J. Wang. Swarm Intelligence. In *Proceedings of the Seventh Annual Meeting of the Robotics Society of Japan*, pages 425–428, Tokyo, 1989.

[CD98] M. Colombetti and M. Dorigo. Evolutionary computation in behavior engineering. In X. Yao, editor, *Evolutionary Computation: Theory and Applications*, pages 37–80, World Scientific, Singapore, 1998.

[CDB96] M. Colombetti, M. Dorigo, and G. Borghi. Behavior analysis and training – A methodology for behavior engineering. *IEEE Transactions on Systems, Man, and Cybernetics – Part B: Cybernetics*, 26(3):365–380, 1996.

[CFK97] Y. U. Cao, A. S. Fukunaga, and A. B. Kahng. Cooperative mobile robotics: Antecedents and directions. *Autonomous Robots*, 4(1):7–27, 1997.

[CFKM95] Y. U. Cao, A. S. Fukunaga, A. B. Kahng, and F. Meng. Cooperative mobile robotics: Antecedents and directions. In *Proceedings of the IEEE/RSJ International Conference on Intelligent Robots and Systems*, pages 226–234, Pittsburgh, 1995.

[CM96] D. Cliff and G. F. Miller. Co-evolution of pursuit and evasion II: Simulation methods and results. In P. Maes et al., editors, *From Animals to Animats 4: Proceedings of the Fourth International Conference on Simulation of Adaptive Behavior*, pages 506–515, The MIT Press, Cambridge, MA, 1996.

[DC94] M. Dorigo and M. Colombetti. Robot shaping: Developing autonomous agents through learning. *Artificial Intelligence*, 71(2):321–370, 1994.

[DGF+91] J. C. Deneubourg, S. Goss, N. Franks, A. Sendova, A. Franks, C. Detrin, and L. Chatier. The dynamics of collective sorting: Robot-like ant and ant-like robot. In J.-A. Meyer and S. Wilson, editors, *From Animals to Animats 1: Proceedings of the First International Conference on Simulation of Adaptive Behavior*, pages 356–365, The MIT Press, Cambridge, MA, 1991.

[DJMW96] G. Dudek, M. Jenkin, E. E. Milios, and D. Wilkes. A taxonomy for multiagent robotics. *Autonomous Robots*, 3(4):375–397, 1996.

[DJR93] B. R. Donald, J. Jennings, and D. Rus. Experimental information invariants for cooperating autonomous mobile robots. In *Proceedings of the International Joint Conference on Artificial Intelligence, Workshop on Dynamically Interacting Robots*, Chambery, 1993.

[DLC89] E. H. Durfee, V. R. Lesser, and D. D. Corkill. Trends in cooperative distributed problem solving. *IEEE Transactions on Knowledge and Data Engineering*, KOE-11(1):63–83, 1989.

[DMC96] M. Dorigo, V. Maniezzo, and A. Colorni. The ant system: Optimization by a colony of cooperating agents. *IEEE Transactions on Systems, Man and Cybernetics*, 26(1):1–13, 1996.

[DN98] K. Dautenhahn and C. Nehaniv. Artificial Life and natural stories. In *Proceedings of the Third International Symposium on Artificial Life and Robotics*, pages 435–439, Beppu, Japan, 1998.

[DS91] M. Dorigo and U. Schnepf. Organization of robot behavior through genetic learning processes. In *Proceedings of the Fifth International Conference on Advanced Robotics*, pages 1456–1460, Pisa, 1991.

[Dye95] M. G. Dyer. Toward synthesizing artificial neural networks that exhibit cooperative intelligent behavior: Some open issues in artificial life. In C. G. Langton, editor, *Artificial Life: An Overview*, pages 111–134, The MIT Press, Cambridge, MA, 1995.

[FAAE98] T. Fujii, Y. Arai, H. Asama, and I. Endo. Multilayered reinforcement learning for complicated collision avoidance problems. In *Proceedings of the IEEE International Conference on Robotics and Automation*, pages 2186–2191, Leuven, Belgium, 1998.

[FAvN+96] T. Fujii, H. Asama, T. von Numers, T. Fujita, H. Kaetsu, and I. Endo. Co-evolution of a multiple autonomous robot system and its working environment via intelligent local information storage. *Robotics and Autonomous Systems*, 19(1):1–13, 1996.

[FBKT99] D. Fox, W. Burgard, H. Kruppa, and S. Thrun. Collaborative multi-robot localization. In *Proceedings of the German Conference on Artificial Intelligence and the 21st Symposium on Pattern Recognition*, pages 255–266, 1999.

[FBKT00] D. Fox, W. Burgard, H. Kruppa, and S. Thrun. A probabilistic approach to collaborative multi-robot localization. *Autonomous Robots*, 8(3):325–344, 2000.

[FFA99] T. Fukuda, D. Funato, and F. Arai. Recognizing environmental change through multiplex reinforcement learning in group robot system. In *Proceedings of the IEEE/RSJ International Conference on Intelligent Robots and Systems*, pages 972–977, Kyongju, Korea, 1999.

[FGM+98] D. Floreano, J. Godjevac, A. Martinoli, F. Mondada, and J.-D. Nicoud. Design, control, and applications of autonomous mobile robots. In S. G. Tzafestas, editor, *Advances in Intelligent Autonomous Agents*, Kluwer Academic Publishers, Boston, 1998.

[FI95] T. Fukuda and G. Iritani. Construction mechanism of group behavior with cooperation. In *Proceedings of the IEEE/RSJ International Conference on Intelligent Robots and Systems*, pages 535–542, Pittsburgh, 1995.

[FM96] D. Floreano and F. Mondada. Evolution of homing navigation in a real mobile robot. *IEEE Transactions on Systems, Man and Cybernetics-Part B: Cybernetics*, 26(3):396–407, 1996.

[FM97] M. Fontan and M. J. Mataric. A study of territoriality: The role of critical mass in adaptive task division. In P. Maes et al., editors, *From Animals to Animats 4: Proceedings of the Fourth International Conference on Simulation of Adaptive Behavior*, pages 553–561, The MIT Press, Cambridge, MA, 1997.

[FM98] D. Floreano and F. Mondada. Hardware solutions for evolutionary robotics. In P. Husbands and J.-A. Meyer, editors, *Proceedings of the First European Workshop on Evolutionary Robotics*, pages 137–151, Springer-Verlag, Berlin, 1998.

[FMSA99] T. Fukuda, H. Mizoguchi, K. Sekiyama, and F. Arai. Group behavior control for MARS (Micro Autonomous Robotic System). In *Proceedings of the IEEE International Conference on Robotics and Automation*, pages 1550–1555, Detroit, 1999.

[FN87] T. Fukuda and S. Nakagawa. A dynamically reconfigurable robotic system. In *Proceedings of the International Conference on Industrial Electronics, Control, and Instrumentation*, pages 588–595, Cambridge, MA, 1987.

[FN97] D. Floreano and S. Nolfi. Adaptive behavior in competitive co-evolutionary robotics. In *Proceedings of the Fourth European Conference on Artificial Life*, pages 378–387, Brighton, 1997.

[FP97] P. Funes and J. Pollack. Computer evolution of buildable objects. In P. Husbands and I. Harvey, editors, *Proceedings of the Fourth European Conference on Artificial Life*, pages 358–367, Brighton, 1997.

[FU98] D. Floreano and J. Urzelai. Evolution and learning in autonomous robotic agents. In T. Mange and M. Tomassini, editors, *Bio-inspired Computing Systems: Towards Novel Computational Architectures*, pages 1–36, Presses Polytechniques et Universitaires Romandes, Lausanne, 1998.

[FU00] D. Floreano and J. Urzelai. Evolutionary robotics: the next generation. In T. Gomi, editor, *Evolutionary Robotics*, Kluwer Academic Publishers, Boston, 2000.

[GM99] D. Goldberg and M. J. Mataric. Coordinating mobile robot group behavior using a model of interaction dynamics. In *Proceedings of the Third International Conference on Autonomous Agents*, pages 100–107, Seattle, 1999.

[Gol89] D. Goldberg. *Genetic Algorithms in Search, Optimization and Machine Learning*. Addison-Wesley, Reading, MA, 1989.

[Har92] I. Harvey. Species adaptation genetic algorithms: The basis for a continuing SAGA. In F. J. Varela and P. Bourgine, editors, *Toward a Practice of Autonomous Systems: Proceedings of the First European Conference on Artificial Life*, pages 346–354, The MIT Press/Bradford Books, Cambridge, MA, 1992.

[Har96] I. Harvey. Artificial evolution and real robots. In M. Sugisaka, editor, *Proceedings of the International Symposium on Artificial Life and Robotics (AROB)*, pages 138–141, Beppu, Japan, 1996.

[Har97] I. Harvey. Artificial evolution for real problems. In T. Gomi, editor, *Evolutionary Robotics: From Intelligent Robots to Artificial Life*, pages 187–220, AAI Books, Ontario, 1997.

[Har00] I. Harvey. Robotics: Philosophy of mind using a screwdriver. In T. Gomi, editor, *Evolutionary Robotics: From Intelligent Robots to Artificial Life*, pages 207–230, AAI Books, Ontario, 2000.

[HBBC96] P. Hebert, S. Betge-Brezetz, and R. Chatila. Decoupling odometry and exteroceptive perception in building a global world map of a mobile robot: The use of local maps. In *Proceedings of the IEEE International Conference on Robotics and Automation*, pages 757–764, Minneapolis, 1996.

[HCH96] P. Husbands, D. Cliff, and I. Harvey. The artificial evolution of robot control systems. In *Proceedings of a Colloquium at The Institution of Mechanical Engineers: Genetic Algorithms in Design Optimisation*, London, 1996.

[HH93] B. A. Huberman and T. Hogg. The emergence of computational ecologies. In L. Nadel and D. Stein, editors, *1992 Lectures in Complex Systems*, Volume V of *SFI Studies in the Sciences of Complexity*, pages 185–205, Addison-Wesley, Reading, 1993.

[HH95] B. A. Huberman and T. Hogg. Communities of practice: Performance and evolution. *Computational and Mathematical Organization Theory*, 1:73–92, 1995.

[HHC92] I. Harvey, P. Husbands, and D. Cliff. Issues in Evolutionary Robotics. Cognitive Science Research Paper 219, School of Cognitive and Computing Sciences, University of Sussex, UK, 1992.

[HHC+96] I. Harvey, P. Husbands, D. Cliff, A. Thompson, and N. Jakobi. Evolutionary robotics at Sussex. In *Proceedings of the International Symposium on Robotics and Manufacturing*, Montpellier, 1996.

[HHCM97] P. Husbands, I. Harvey, D. Cliff, and G. Miller. Artificial evolution: A new path for artificial intelligence? *Brain and Cognition*, 34:130–159, 1997.

[HKA+99] Y. Hirata, K. Kosuge, H. Asama, H. Kaetsu, and K. Kawabata. Decentralized control of mobile robots in coordination. In *Proceedings of the IEEE International Conference on Control Applications*, pages 1129–1134, Hilo, 1999.

[Hol75] J. H. Holland. *Adaptation in Natural and Artificial Systems.* The University of Michigan Press, Ann Arbor, 1975.

[Hor99] W. Hordijk. Dynamics, Emergent Computation, and Evolution in Cellular Automata. Ph.D. Thesis, University of New Mexico, USA, 1999.

[Hus98] P. Husbands. Evolving robot behaviors with diffusing gas networks. In P. Husbands and J.-A. Meyer, editors, *Proceedings of the First European Workshop on Evolutionary Robotics*, pages 71–86, Springer-Verlag, Berlin, 1998.

[IKW+97] A. Ishiguro, T. Kondo, Y. Watanabe, Y. Shirai, and Y. Uchikawa. Emergent construction of artificial immune networks for autonomous robots. In *Proceedings of the IEEE International Conference on Systems, Man and Cybernetics*, pages 1222–1228, Orlando, 1997.

[IOH98] K. Inoue, J. Ota, and T. Hirano. Interactive transportation by cooperative mobile robots in unknown environment. In T. Lueth et al., editors, *Distributed Autonomous Robotic Systems 3*, pages 3–12, Springer-Verlag, Berlin, 1998.

[Jak98a] N. Jakobi. The minimal simulation approach to evolutionary robotics. In T. Gomi, editor, *Evolutionary Robotics: From Intelligent Robots to Artificial Life*, AAI Books, Ontario, 1998.

[Jak98b] N. Jakobi. Minimal simulations for evolutionary robotics. D.Phil. Thesis, University of Sussex, UK, 1998.

[JD87] V. Jagannathan and R. Dodhiawak. Distributed artificial intelligence: An automated bibliography. In M. Huhns, editor, *Distributed Artificial Intelligence*, pages 341–390, Morgan Kaufmann Publishers, 1987.

[JGC+97] J. A. Janet, R. Gutierrez, T. A. Chase, M. W. White, and J. C. Sutton. Autonomous mobile robot global self-localization using Kohonen and region-feature neural networks. *Journal of Robotic Systems*, 14(4):263–282, 1997.

[JQ98] N. Jakobi and M. Quinn. Some problems (and a few solutions) for open-ended evolutionary robotics. In P. Husbands and J.-A. Meyer, editors, *Proceedings of the First European Workshop on Evolutionary Robotics*, pages 108–122, Springer-Verlag, Berlin, 1998.

[KE94] B. J. A. Krose and M. Eecen. A self-organizing representation of sensor space for mobile robot navigation. In *Proceedings of the IEEE/RSJ International Conference on Intelligent Robots and Systems*, pages 9–14, Munich, 1994.

[Kha85] O. Khatib. Real-time obstacle avoidance for manipulators and mobile robots. In *Proceedings of the IEEE International Conference on Robotics and Automation*, pages 500–505, St. Louis, 1985.

[Kha86] O. Khatib. Real-time obstacle avoidance for manipulators and mobile robots. *The International Journal of Robotics Research*, 5(1):90–98, 1986.

[Kha87] O. Khatib. A unified approach to motion and force control of robot manipulators: The operation space formulation. *IEEE Journal of Robotics and Automation*, 3(1):43–53, 1987.

[KK95] A. A. Kassim and B. V. K. V. Kumar. Potential fields and neural networks. In M. A. Arbib, editor, *The Handbook of Brain Theory and Neural Networks*, pages 749–753, The MIT Press, Cambridge, MA, 1995.

[KLM96] L. P. Kaelbling, M. L. Littman, and A. W. Moore. Reinforcement learning: A survey. *Journal of Artificial Intelligence Research*, 4:237–285, 1996.

[Koh88] T. Kohonen. *Self-organization and Associative Memory*. Springer-Verlag, New York, 1988.

[Koz92] J. R. Koza. *Genetic Programming: On the Programming of Computers by Means of Natural Selection*. The MIT Press, Cambridge, MA, 1992.

[KR92] J. R. Koza and J. P. Rice. Automatic programming of robots using genetic programming. In *Proceedings of the Tenth National Conference on Artificial Intelligence*, pages 194–207, San Jose, 1992.

[KS60] J. G. Kemeny and J. L. Snell. *Finite Markov Chains*. Springer-Verlag, New York, 1960.

[KSO+97] J. H. Kim, I. H. Suh, S. R. Oh, Y. J. Cho, and Y. K. Chung. Region-based Q-learning using convex clustering approach. In *Proceedings of the IEEE/RSJ International Conference on Intelligent Robots and Systems*, pages 601–607, Grenoble, 1997.

[KZ97] C. R. Kube and H. Zhang. Task modeling in collective robotics. *Autonomous Robots*, 4:53–72, 1997.

[Lan88] C. G. Langton. Artificial life. In *Artificial Life: Proceedings of an Interdisciplinary Workshop on the Synthesis and Simulation of Living Systems*, pages 1–47, Los Alamos, 1988.

[Lat91] J.-C. Latombe. *Robot Motion Planning*. Kluwer, Norwell, MA, 1991.

[LDK95] M. L. Littman, T. L. Dean, and L. Kaelbling. On the complexity of solving Markov decision problems. In *Proceedings of the Eleventh Annual Conference on Uncertainty in Artificial Intelligence*, Volume 11, pages 394–402, Montreal, 1995.

[LFB94] M. Lewis, A. Fagg, and G. Bekey. Genetic algorithms for gait synthesis in a hexapod robot. In Y. Zheng, editor, *Recent Trends in Mobile Robots*, pages 317–331, World Scientific, Singapore, 1994.

[LH94] D. M. Lyons and A. J. Hendriks. Testing incremental adaptation. In *Proceedings of the Second International Conference on AI Planning Systems*, pages 116–121, Chicago, 1994.

[LHF+97] S. Luke, C. Hohn, J. Farris, G. Jackson, and J. Hendler. Co-evolving soccer softbot team coordination with genetic programming. In *Proceedings of the RoboCup-97 Workshop, the 15th International Joint Conference on Artificial Intelligence*, pages 115–118, Nagoya, 1997.

[LHL97a] W.-P. Lee, J. Hallam, and H. H. Lund. Applying genetic programming to evolve behavior primitives and arbitrators for mobile robots. In *Proceedings of the IEEE Fourth International Conference on Evolutionary Computation*, pages 501–506, IEEE Press, Piscataway, 1997.

[LHL97b] H. H. Lund, J. Hallam, and W.-P. Lee. Evolving robot morphology. In *Proceedings of the IEEE Fourth International Conference on Evolutionary Computation*, IEEE Press, Piscataway, 1997.

[Lin93] L.-J. Lin. Hierarchical learning of robot skills by reinforcement. In *Proceedings of the International Conference on Neural Networks*, pages 181–186, San Francisco, 1993.

[LK97] S. Lee and G. Kardaras. Elastic string based global path planning using neural networks. In *Proceedings of the IEEE International Symposium on Computational Intelligence in Robotics and Automation*, pages 108–114, Monterey, 1997.

[LM92] L.-J. Lin and T. M. Mitchell. Reinforcement learning with hidden states. In J.-A. Meyer, H. L. Roitblat, and S. Wilson, editors, *From Animals to Animats 2: Proceedings of the Second International Conference on the Simulation of Adaptive Behavior*, pages 271–280, The MIT Press, Cambridge, MA, 1992.

[LVCS93] G. Lucarini, M. Varioli, R. Cerutti, and G. Sandini. Cellular robotics: Simulation and HW implementation. In *Proceedings of the IEEE International Conference on Robotics and Automation*, Volume 3, pages 846–852, IEEE Press, Piscataway, 1993.

[MAAO+99] S. Marsella, J. Adibi, Y. Al-Onaizan, G. Kaminka, I. Muslea, and M. Tambe. On being a teammate: Experiences acquired in the design of RoboCup teams. In O. Etzioni, J. Muller, and J. Bradshaw, editors, *Proceedings of the Third International Conference on Autonomous Agents*, pages 221–227, Seattle, 1999.

[Mae89] P. Maes. The dynamics of action selection. In *Proceedings of the International Joint Conference on Artificial Intelligence*, pages 991–997, Detroit, 1989.

[Mae95] P. Maes. Modeling adaptive autonomous agents. In C. G. Langton, editor, *Artificial Life: An Overview*, pages 135–162, The MIT Press, Cambridge, MA, 1995.

[Mat92a] M. J. Mataric. Behavior-based systems: Key properties and implications. In *Proceedings of the Workshop on Architectures for Intelligent Control Systems, IEEE International Conference on Robotics and Automation*, pages 46–54, Nice, 1992.

[Mat92b] M. J. Mataric. Designing emergent behaviors: From local interactions to collective intelligence. In J.-A. Meyer, H. L. Roitblat, and S. Wilson, editors, *From Animals to Animats 2: Proceedings of the Second International Conference on Simulation of Adaptive Behavior*, pages 432–441, The MIT Press, Cambridge, MA, 1992.

[Mat92c] M. J. Mataric. Minimizing complexity in controlling a mobile robot population. In *Proceedings of the IEEE International Conference on Robotics and Automation*, pages 830–835, Nice, 1992.

[Mat93] M. J. Mataric. Synthesizing group behaviors. In *Proceedings of the Workshop on Dynamically Interacting Robots, International Joint Conference on Artificial Intelligence*, pages 1–10, Chambery, 1993.

[Mat94a] M. J. Mataric. Interaction and intelligent behavior. Ph.D. Thesis, Department of Electrical Engineering and Computer Science, MIT, USA, 1994.

[Mat94b] M. J. Mataric. Learning motor skills by imitation. In *Toward Physical Interaction and Manipulation: Proceedings of AAAI Spring Symposium*, Stanford University, 1994.

[Mat94c] M. J. Mataric. Learning to behave socially. In D. Cliff, P. Husbands, J.-A. Meyer, and S. Wilson, editors, *From Animals to Animats 3: Proceedings of the Third International Conference on Simulation of Adaptive Behavior (SAB-94)*, pages 453–462, The MIT Press, Cambridge, MA, 1994.

[Mat94d] M. J. Mataric. Reward functions for accelerated learning. In W. W. Cohen and H. Hirsh, editors, *Proceedings of the Eleventh International Conference on Machine Learning*, Morgan Kaufmann Publishers, San Francisco, 1994.

[Mat95a] M. J. Mataric. Designing and understanding adaptive group behavior. *Adaptive Behavior*, 4(1):51–80, 1995.

[Mat95b] M. J. Mataric. Issues and approaches in the design of collective autonomous agents. *Robotics and Autonomous Systems*, 16(2–4):321–331, 1995.

[Mat96] M. J. Mataric. Learning in multi-robot systems. In G. Weiss and S. Sen, editors, *Adaptation and Learning in Multi-Agent Systems, Lecture Notes In Artificial Intelligence, Vol. 1042*, pages 152–163, Springer-Verlag, Berlin, 1996.

[Mat97a] M. J. Mataric. Learning social behavior. *Robotics and Autonomous Systems*, 20:191–204, 1997.

[Mat97b] M. J. Mataric. Reinforcement learning in the multi-robot domain. *Autonomous Robots*, 4(1):73–83, 1997.

[Mat98] M. J. Mataric. Behavior-based robotics as a tool for synthesis of artificial behavior and analysis of natural behavior. *Trends in Cognitive Science*, 2(3):82–87, 1998.

[MB90] P. Maes and R. A. Brooks. Learning to coordinate behaviors. In *Proceedings of the Eighth National Conference on Artificial Intelligence*, pages 796–802, Boston, 1990.

[MB93] D. McFarland and U. Bosser. *Intelligent Behavior in Animals and Robots*. The MIT Press, Cambridge, MA, 1993.

[MC92] S. Mahadevan and J. Connell. Automatic programming of behavior-based robots using reinforcement learning. *Artificial Intelligence*, 55(2-3):311–365, 1992.

[MC96] M. J. Mataric and D. Cliff. Challenges in evolving controllers for physical robots. *Evolutional Robotics, Special Issue of Robotics and Autonomous Systems*, 19(1):67–83, 1996.

[McF94] D. McFarland. Towards robot cooperation. In D. Cliff, P. Husbands, J.-A. Meyer, and S. Wilson, editors, *From Animals to Animats 3: Proceedings of the Third International Conference on Simulation of Adaptive Behavior*, pages 440–444, The MIT Press, Cambridge, MA, 1994.

[MFA+96] N. Mitsumoto, T. Fukuda, F. Arai, H. Tadashi, and T. Idogaki. Self-organizing multiple robotic system. In *Proceedings of the IEEE International Conference on Robotics and Automation*, pages 1614–1619, Minneapolis, 1996.

[MFM97] A. Martinoli, E. Franzi, and O. Matthey. Towards a reliable set-up for bio-inspired collective experiments with real robots. In A. Casals and A. T. de Almeida, editors, *Proceedings of the Fifth International Symposium on Experimental Robotics, Lecture Notes in Control and Information Sciences*, pages 597–608, Springer-Verlag, Berlin, 1997.

[Mic92] Z. Michalewicz. *Genetic Algorithms + Data Structures = Evolution Programs*. Springer-Verlag, Berlin, 1992.

[ML95] A. Murray and S. J. Louis. Design strategies for evolutionary robotics. In E. A. Yfantis, editor, *Intelligent Systems: Proceedings of the Third Golden West International Conference*, pages 609–616, Kluwer Academic Publishers, Norwell, MA, 1995.

[MLN96] O. Miglino, H. H. Lund, and S. Nolfi. Evolving mobile robots in simulated and real environments. *Artificial Life*, 2:417–434, 1996.

[MM95] A. Martinoli and F. Mondada. Collective and cooperative behaviors: Biologically inspired experiments in robotics. In *Proceedings of the Fourth International Symposium on Experimental Robotics*, pages 2–7, Stanford University, 1995.

[MMHM00] B. Minten, R. Murphy, J. Hyams, and M. Micire. A communication-free behavior for docking mobile robots. In L. E. Parker, G. Bekey, and J. Barhen, editors, *Distributed Autonomous Robotic Systems 4*, Springer-Verlag, New York, 2000.

[MNS95] M. J. Mataric, M. Nilsson, and K. Simsarian. Cooperative multi-robot box-pushing. In *Proceedings of the IEEE/RSJ International Conference on Intelligent Robots and Systems*, pages 556–561, Pittsburgh, 1995.

[MOA+00] N. Miyata, J. Ota, Y. Aiyama, H. Asama, and T. Arai. Cooperative transport in unknown environment — Application of real-time task assignment. In *Proceedings of the 2000 IEEE International Conference on Robotics and Automation*, pages 3176–3182, San Francisco, 2000.

[Mor88] H. Moravec. *Mind Children: The Future of Robot and Human Intelligence*. Harvard University Press, Cambridge, MA, 1988.

[NF98] S. Nolfi and D. Floreano. How co-evolution can enhance the adaptive power of artificial evolution: Implications for evolutionary robotics. In P. Husbands and J.-A. Meyer, editors, *Proceedings of the First European Workshop on Evolutionary Robotics*, pages 22–38, Springer-Verlag, Berlin, 1998.

[NF99] S. Nolfi and D. Floreano. Learning and evolution. *Autonomous Robots*, 7(1):89–113, 1999.

[NFMM94] S. Nolfi, D. Floreano, O. Miglino, and F. Mondada. How to evolve autonomous robots: Different approaches in evolutionary robotics. In R. A. Brooks and P. Maes, editors, *Proceedings of the Fourth International Conference on Artificial Life*, pages 190–197, The MIT Press, Cambridge, MA, 1994.

[NID00] Y. Nagayuki, S. Ishii, and K. Doya. Multi-agent reinforcement learning: An approach based on the other agent's internal model. In *Proceedings of the Fourth International Conference on Multi-Agent Systems*, pages 215–221, Boston, 2000.

[NNS97] H. F. Nijhout, L. Nadel, and D. L. Stein. *Pattern Formation in the Physical and Biological Sciences*. Addison-Wesley, Reading, MA, 1997.

[Nol98] S. Nolfi. Evolutionary robotics: Exploiting the full power of self-organization. *Connection Science*, 10(3–4):167–183, 1998.

[NP95] S. Nolfi and D. Parisi. Evolving non-trivial behaviors on real robots: An autonomous robot that picks up objects. In M. Gori and E. Soda, editors, *Proceedings of the Fourth International Congress of the Italian Association of Artificial Intelligence*, Springer-Verlag, Berlin, 1995.

[NV92] A.E. Nix and M.D. Vose. Modeling genetic algorithms with markov chains. *Annals of Mathematics and Artificial Intelligence*, 5(1):79–88, 1992.

[OJ96] G. M. P. O'Hare and N. R. Jennings. *Foundations of Distributed Artificial Intelligence*. John Wiley and Sons, Inc., New York, 1996.

[Par94] L. E. Parker. Heterogeneous Multi-Robot Cooperation. Ph.D. Thesis, MIT, USA, 1994.

[Par95] L. E. Parker. The effect of action recognition and robot awareness in cooperative robotic teams. In *Proceedings of the IEEE/RSJ International Conference on Intelligent Robots and Systems*, pages 212–219, Pittsburgh, 1995.

[Par97] D. Parisi. Artificial life and higher level cognition. *Brain and Cognition*, 34:160–184, 1997.

[Par99] L. E. Parker. Cooperative robotics for multi-target observation. *Intelligent Automation and Soft Computing*, 5(1):5–19, 1999.

[Par00] L. E. Parker. Current state of the art in distributed robot systems. In L. E. Parker, G. Bekey, and J. Barhen, editors, *Distributed Autonomous Robotic Systems 4*, pages 3–12, Springer-Verlag, Berlin, 2000.

[Per92] A. S. Perelson. *Theoretical Immunology*. Addison-Wesley, Reading, MA, 1992.

[Pfe98] R. Pfeifer. Embodied system life. In *Proceedings of the International Symposium on System Life*, Tokyo, 1998.

[PM00] P. Pirjanian and M. J. Mataric. Multi-robot target acquisition using multiple objective behavior coordination. In *Proceedings of the IEEE International Conference on Robotics and Automation*, San Francisco, 2000.

[PN96] D. Parisi and S. Nolfi. The influence of learning on evolution. In R. K. Belew and M. Mitchell, editors, *Adaptive Individuals in Evolving Populations*, pages 419–428, Addison-Wesley, Reading, MA, 1996.

[Pra95] E. Prassler. Robot navigation: A simple guidance system for a complex changing world. In H. Bunke, T. Kanade, and H. Noltemeier, editors, *Modeling and Planning for Sensor Based Intelligent Robot Systems*, pages 86–103, World Scientific, Singapore, 1995.

[PS98] R. Pfeifer and C. Scheier. Embodied cognitive science: A novel approach to the study of intelligence in natural and artificial systems. In T. Gomi, editor, *Evolutionary Robotics: From Intelligent Robots to Artificial Life*, pages 1–35, AAI Books, Ontario, 1998.

[PT00] L. E. Parker and C. Touzet. Multi-robot learning in a cooperative observation task. In L. E. Parker, G. Bekey, and J. Barhen, editors, *Distributed Autonomous Robotic Systems 4*, pages 391–401, Springer-Verlag, Berlin, 2000.

[PY90] S. Premvuti and S. Yuta. Consideration on the cooperation of multiple autonomous mobile robots. In *Proceedings of the IEEE International Workshop on Intelligent Robots and Systems*, pages 59–63, Tsuchiura, 1990.

[QK93] S. Quinlan and O. Khatib. Elastic bands: Connecting path planning and control. In *Proceedings of the IEEE International Conference on Robotics and Automation*, Volume 2, pages 802–807, Atlanta, 1993.

[RABP94] A. Ram, R. C. Arkin, G. Boone, and M. Pearce. Using genetic algorithms to learn reactive control parameters for autonomous robotic navigation. *Adaptive Behavior*, 2(3):277–304, 1994.

[Rey92] C. Reynolds. An evolved, vision-based behavioral model of coordinated group motion. In J.-A. Meyer, H. L. Roitblat, and S. Wilson, editors, *From Animals to Animats 2: Proceedings of the Second International Conference on Simulation of Adaptive Behavior*, pages 384–392, The MIT Press, Cambridge, MA, 1992.

[Rey94] C. Reynolds. Evolution of corridor following in a noisy world. In D. Cliff, P. Husbands, J.-A. Meyer, and S. Wilson, editors, *From Animals to Animats 3: Proceedings of the Third International Conference on Simulation of Adaptive Behavior*, pages 402–410, The MIT Press/Bradford Books, Cambridge, MA, 1994.

[Ros93] J. S. Rosenschein. Consenting agents: Negotiation mechanisms for multi-agent systems. In *Proceedings of the International Joint Conference on Artificial Intelligence*, pages 792–799, Chambery, 1993.

[RV00] P. Riley and M. Veloso. On behavior classification in adversarial environments. In L. E. Parker, G. Bekey, and J. Barhen, editors, *Distributed Autonomous Robotic Systems 4*, Springer-Verlag, New York, 2000.

[SAB+00] R. Simmons, D. Apfelbaum, W. Burgard, D. Fox, M. Moors, S. Thrun, and H. Younes. Coordination for multi-robot exploration and mapping. In *Proceedings of the National Conference on Artificial Intelligence*, Austin, 2000.

[SB93] D. J. Stilwell and J. S. Bay. Toward the development of a material transport system using swarms of ant-like robots. In *Proceedings of the IEEE International Conference on Robotics and Automation*, pages 766–771, Atlanta, 1993.

[SB98] R. S. Sutton and A. G. Barto. *Reinforcement Learning: An Introduction*. The MIT Press, Cambridge, MA, 1998.

[Sch94] A. C. Schultz. Learning robot behaviors using genetic algorithms. In M. Jamshidi and C. Nguyen, editors, *Proceedings of the First World Automation Congress*, pages 607–612, TSI Press, Albuquerque, 1994.

[Sch95] H.-P. Schwefel. *Evolution and Optimum Seeking*. John Wiley and Sons, Inc., New York, 1995.

[Sch99] S. Schaal. Is imitation learning the route to humanoid robots? *Trends in Cognitive Sciences*, 3:233–242, 1999.

[Sch00] S. Schaal. Robot learning. In M. A. Arbib, editor, *The Handbook of Brain Theory and Neural Networks*, The MIT Press, Cambridge, MA, 2000.

[SG94] A. C. Schultz and J. Grefenstette. Evolving robot behaviors. In *Proceedings of the Fourth International Workshop on the Synthesis and Simulation of Living Systems*, Boston, 1994.

[SH94] L. Spector and J. Hendler. The use of supervenience in dynamic-world planning. In *Proceedings of the Second International Conference on AI Planning Systems*, pages 158–163, Chicago, 1994.

[Sha94] M. Shanahan. Evolutionary automata. In R. A. Brooks and P. Maes, editors, *Artificial Life IV: Proceedings of the Fourth International Workshop on the Synthesis and Simulation of Living Systems*, pages 387–393, The MIT Press, Cambridge, MA, 1994.

[Sha97] N. E. Sharkey. The new wave in robot learning. *Robotics and Autonomous Systems*, 22:179–186, 1997.

[Sim94] K. Sims. Evolving 3D morphology and behavior by competition. In R. A. Brooks and P. Maes, editors, *Artificial Life IV: Proceedings of the Fourth International Workshop on the Synthesis and Simulation of Living Systems*, pages 28–39, The MIT Press, Cambridge, MA, 1994.

[SL93] T. Sugawara and V. Lesser. Learning coordination plans in distributed problem-solving environments. Computer Science Technical Report 93-27, University of Massachusetts, USA, 1993.

[SM00] G. S. Sukhatme and M. J. Mataric. Embedding robots into the Internet. In D. Heidamann, editor, *Communications of the ACM – Special Issue on Embedding the Internet*, Volume 43, pages 67–73, 2000.

[SP96] G. Saunders and J. B. Pollack. The evolution of communication schemes of continuous channels. In P. Maes, M. J. Mataric, J.-A. Meyer, J. B. Pollack, and S. Wilson, editors, *From Animals to Animats 4: Proceedings of the Fourth International Conference on Simulation of Adaptive Behavior*, pages 580–589, The MIT Press, Cambridge, MA, 1996.

[Spi97] E. Spier. From reactive behavior to adaptive behavior: Motivational models for behavior in animals and robots. D.Phil. Thesis, Oxford University, UK, 1997.

[SSH94] S. Sen, M. Sekaran, and J. Hale. Learning to coordinate without sharing information. In *Proceedings of the National Conference on Artificial Intelligence*, pages 426–431, Seattle, 1994.

[Ste90] L. Steels. Exploiting analogical representations. In P. Maes, editor, *Designing Autonomous Agents*, pages 71–88, The MIT Press, Cambridge, MA, 1990.

[Ste94] L. Steels. Emergent functionality through on-line evolution. In R. A. Brooks and P. Maes, editors, *Artificial Life IV:Proceedings of the Fourth International Workshop on the Synthesis and Simulation of Living Systems*, pages 8–14, The MIT Press, Cambridge, MA, 1994.

[Ste95] L. Steels. The artificial life roots of artificial intelligence. In C. G. Langton, editor, *Artificial Life: An Overview*, pages 75–110, The MIT Press, Cambridge, MA, 1995.

[Sut88] R. S. Sutton. Learning to predict by method of temporal differences. *Journal of Machine Learning*, 3(1):9–44, 1988.

[SV97] P. Stone and M. Veloso. Multiagent Systems: A Survey from a Machine Learning Perspective. Technical Report CMU-CS-97-193, School of Computer Science, Carnegie Mellon University, USA, 1997.

[SV98] P. Stone and M. Veloso. A layered approach to learning client behaviors in the RoboCup soccer server. *Applied Artificial Intelligence*, 12(2-3):165–188, 1998.

[TA99] Y. Takahashi and M. Asada. Behavior acquisition by multi-layered reinforcement learning. In *Proceedings of the IEEE International Conference on Systems, Man, and Cybernetics*, pages 716 - 721, Tokyo, 1999.

[TFB98] S. Thrun, D. Fox, and W. Burgard. A probabilistic approach to concurrent mapping and localization for mobile robots. *Machine Learning and Autonomous Robots (joint issue)*, 31(5):1–25, 1998.

[Thr00] S. Thrun. Probabilistic Algorithms in Robotics. Technical Report CMU-CS-00-126, School of Computer Science, Carnegie Mellon University, USA, 2000.

[UAH98] E. Uchibe, M. Asada, and K. Hosoda. Cooperative behavior acquisition in multi-mobile robots environment by reinforcement learning based on state vector estimation. In *Proceedings of the IEEE International Conference on Robotics and Automation*, pages 1558–1563, Leuven, Belgium, 1998.

[UF00] J. Urzelai and D. Floreano. Evolutionary robots with fast adaptive behavior in new environments. In T. C. Fogarty, J. Miller, A. Thompson, and P. Thomson, editors, *From Biology to Hardware: Proceedings of the Third International Conference on Evolvable Systems*, Springer-Verlag, Berlin, 2000.

[UNA98] E. Uchibe, M. Nakamura, and M. Asada. Co-evolution for cooperative behavior acquisition in a multiple mobile robot environment. In *Proceedings of the IEEE/RSJ International Conference on Intelligent Robots and Systems*, pages 425–430, Victoria, BC, 1998.

[VBX96] J. Vandorpe, H. Van Brussel, and H. Xu. Exact dynamic map building for a mobile robot using geometrical primitives produced by a 2D ranger finder. In *Proceedings of the IEEE International Conference on Robotics and Automation*, pages 901–908, Minneapolis, 1996.

[Ven94] J. Ventrella. Explorations in the emergence of morphology and locomotion behavior in animated characters. In R. A. Brooks and P. Maes, editors, *Artificial Life IV: Proceedings of the Fourth International Workshop on the Synthesis and Simulation of Living Systems*, pages 436–441, The MIT Press, Cambridge, MA, 1994.

[VG97] C. Versino and L. M. Gambardella. Ibots learn genuine team solutions. In M. Van Someren and G. Widmer, editors, *Proceedings of the European Conference on Machine Learning, Lecture Notes in Artificial Intelligence, Vol. 1224*, pages 298–311, Springer-Verlag, Berlin, 1997.

[Wan89] P. K. C. Wang. Navigation strategies for multiple autonomous mobile robots. In *Proceedings of the IEEE/RSJ International Workshop on Intelligent Robots and Systems*, pages 486–493, Tsukuba, 1989.

[WB90] S. D. Whitehead and D. Ballard. Learning to Perceive and Act. Technical Report TR-331, Department of Computer Science, University of Rochester, USA, 1990.

[WD92] C. Watkins and P. Dayan. Technical notes: Q-learning. *Machine Learning*, 8:279–292, 1992.

[WD99] G. Weiss and P. Dillenbourg. What is 'multi' in multiagent learning? In P. Dillenbourg, editor, *Collaborative Learning. Cognitive and Computational Approaches*, pages 64–80, Pergamon Press, Oxford, 1999.

[Wei96] G. Weiss. Adaptation and learning in multi-agent systems: Some remarks and a bibliography. In G. Weiss and S. Sen, editors, *Adaptation and Learning in Multi-Agent Systems, Lecture Notes in Artificial Intelligence, Vol. 1042*, pages 1–21, Springer-Verlag, Berlin, 1996.

[Wei99] G. Weiss. *Multiagent Systems: A Modern Approach to Distributed Artificial Intelligence*. The MIT Press, Cambridge, MA, 1999.

[WFP00] R. A. Watson, S. G. Ficici, and J. B. Pollack. Embodied Evolution: Distributing an Evolutionary Algorithm in a Population of Robots. Technical Report CS-00-208, Volen Center for Complex Systems, Brandeis University, USA, 2000.

[Win90] A. T. Winfree. *The Geometry of Biological Time*. Springer-Verlag, New York, 1990.

[WJ95] M. Wooldridge and N. R. Jennings. Intelligent agents: Theory and practice. *The Knowledge Engineering Review*, 10(2):115–152, 1995.

[WT99] D. Wolpert and K. Tumer. An Introduction to Collective Intelligence. Technical Report NASA-ARC-IC-99-63, NASA Ames Research Center, USA, 1999.

[Wya97] J. Wyatt. Exploration and Inference in Learning from Reinforcement. Ph.D. Thesis, Department of Artificial Intelligence, University of Edinburgh, UK, 1997.

[XMZT97] J. Xiao, Z. Michalewicz, L. Zhang, and K. Trojanowski. Adaptive evolutionary planner/navigator for mobile robots. *IEEE Transaction on Evolutionary Computation*, 1(1):18–28, 1997.

[YFO+00] A. Yamashita, M. Fukuchi, J. Ota, T. Arai, and H. Asama. Motion planning for cooperative transportation of a large object by multiple mobile robots in a 3D environment. In *Proceedings of the IEEE International Conference on Robotics and Automation*, pages 3144–3151, San Francisco, 2000.

[YH95] M. Youssefmir and B. A. Huberman. Clustered Volatility in Multiagent Systems. Technical Report, XEROX PARC, USA, 1995.

[YTY97] T. Yamaguchi, Y. Tanaka, and M. Yachida. Speed up reinforcement learning between two agents with adaptive mimetism. In *Proceedings of the IEEE/RSJ International Conference on Intelligent Robots and Systems*, pages 594–600, Grenoble, 1997.

[Zie98] T. Ziemke. Adaptive behavior in autonomous agents. *Presence*, 7(6):564–587, 1998.

Index